DYNAMICS OF HIGH-SPEED RAILWAY BRIDGES

Dynamics of High-Speed Railway Bridges

Editors

Raimundo Delgado & Rui Calçada
Faculty of Engineering, Department of Civil Engineering,
University of Porto, Portugal

José Maria Goicolea & Felipe Gabaldón
Computational Mechanics Group, Department of Mechanics and Structures,
E.T.S. Ingenieros de Caminos, Universidad Politécnica de Madrid, Spain

CRC Press
Taylor & Francis Group
Boca Raton London New York

CRC Press is an imprint of the
Taylor & Francis Group, an informa business
A BALKEMA BOOK

Universidade do Porto
Faculdade de Engenharia
FEUP

ETSICCyP

Published by:
CRC Press/Balkema
Schipholweg 107C, 2316 XC Leiden, The Netherlands

ISBN 13: 978-0-415-46767-4 (hbk)

Typeset by Charon Tec Ltd (A Macmillan Company), Chennai, India

Library of Congress Cataloging-in-Publication Data

Dynamics of high-speed railway bridges / Raimundo Delgado … [et al.].
p. cm.
Includes bibliographical references and index.
ISBN 978-0-415-46767-4 (hardcover : alk. paper)—ISBN 978-0-203-89540-5 (ebook : alk. paper)
1. High speed trains. 2. Railroad bridges. 3. Bridges–Live loads. 4. Dynamics.
I. Delgado, Raimundo.

TF1460.D96 2008
625.1–dc22

2008003169

Table of Contents

Preface

The construction of high-speed railways comprises a set of demands, from safety aspects to new types of equipment and construction solutions, involving the most recent and sophisticated technologies.

Among these, emphasis is given to the railway behaviour where the structural elements are of great relevance. The dynamics of the structure is in this case, more than in others, controlling the process, being a crucial factor in issues such as the structural and railway safety and the passengers comfort.

The ability to address the multiple issues relevant to this process requires expertise and know-how on the dynamic analysis techniques, which have been recently developed in this field, with repercussions in terms of the European regulations in this domain.

The themes included in this book are mainly based on the papers presented at the advanced course and workshop "DYNAMICS OF HIGH-SPEED RAILWAY BRIDGES" organised by the Faculdade de Engenharia da Universidade do Porto (FEUP) and the Escuela Tecnica Superior de Ingenieros de Caminos Canales y Puertos de Madrid (ETSICCyP). This book is included in a set of three books: one with a more general thematic "BRIDGES FOR HIGH-SPEED RAILWAYS" and other with a more focused thematic, such as the present book, "TRACK-BRIDGE INTERACTION ON HIGH-SPEED RAILWAYS".

The editors would like to thank all those who contributed to this book, in particular our distinguished guest chapters' authors who heightened, with their knowledge and expertise, to the present interest and quality of the book, the support of the sponsors for the events which originated the materials for this book, and the institutional support of the Faculty of Engineering of the University of Porto and the RAVE – Rede Ferroviária de Alta Velocidade, S.A.

We hope this book will be helpful not only to those professionals involved in the design, construction or maintenance of high speed railway systems, but also to researchers and students working in this field.

List of Authors

Diogo Ribeiro, *FEUP – Portugal*
Francisco Riquelme, *UPM – Spain*
Felipe Gabaldón, *UPM – Spain*
Helder Figueiredo, *FEUP – Portugal*
Ilídio Faria, *UTAD – Portugal*
J. Huyard, *Systra – France*
Juan José Arribas, *UPM – Spain*
José Maria Goicolea-Ruigómez, *UPM – Spain*
José Rui Pinto, *FEUP – Portugal*
Ladislav Frýba, *ITAM – Czech Republic*
Manfred Zacher, *Deutsche Bahn AG – Germany*
Matthias Baeβler, *Federal Research Institute – Germany*
Paul Norris, *Mott MacDonald – United Kingdom*
Philippe Ramondenc, *SNCF – France*
Philippe Van Bogaert, *Ghent University – Belgium*
Raimundo Delgado, *FEUP – Portugal*
Rui Calçada, *FEUP – Portugal*
Serge Montens, *Systra – France*
Wasoodev Hoorpah, *MIO – France*

CHAPTER 1

Dynamic behaviour of rail bridges: The train excitation

P. Ramondenc
SNCF Engineering Department, Paris, France

ABSTRACT: From experience, the control of the dynamic behaviour of rail bridges (deflection and acceleration) has proven to be a "since qua non" condition for the control of railway high-speed. Without it, not only the rail-wheel contact might be ruptured when crossing the structures, but above all chances are that instabilities of the ballast railway might appear on and around the bridges and viaducts. The analysis of the behaviour of the latter using finite-element software allows avoiding such risks. Yet the use of computerized programs does not truly permit to underline the importance of the parameters that rule the phenomenon. Also, engineers do not have at their disposal all the elements permitting to guarantee a real conception: to understand why and how the structure reacts to the crossing of the trains, to know on which characteristic of the structure one has to work in order to modify its behaviour – or simply to foresee the latter (the behaviour) in the case of the degradation of one of his characteristics. Even several numerical simulations, may they be long and expensive, do not allow the total understanding of the train/structure interaction. When analyzing the excitation created by a train, it is possible to underline all the parameters that are useful to the control of the dynamic behaviour of railway bridges. The method presented in this work and based on the concept of the dynamic signature of trains (or their spectrum) allowed elaborating new regulations that were incorporated into the Eurocodes – in particular the universal dynamic train – as well as interoperability technical specifications (ITS) permitting high-speed trains to travel all over Europe. However, the knowledge of those different regulations is not enough to fully understand the phenomenon; only the process described in the present document permits to obtain such a result.

1 GENERAL

Designing railway bridges requires the checking of two main criteria:

a) Resistance and structural durability (generally through ultimate limit state criteria).
b) Serviceability (serviceability limit states) that can also be divided into two criteria
   - Safety of train operations on the bridge, that requires a sufficient track stability and a permanent contact between rail and wheels;
   - Comfort inside vehicles.

All these verifications depend upon the dynamic behaviour of the bridge, not only in terms of deformations (deflections) or stresses but also in terms of level of vibrations (acceleration). Checking accelerations is unusual and need dynamic analysis.

For many years dynamic effects have been taken into account through dynamic coefficients or impact factors leading to static amplified stresses or deformations equivalent to maximum dynamic ones.

This impact factor was determined when solving the equation of vibration of a beam (the well-known partial differential equation), of which the second term is the following excitation function:

$$P(x,t) = \delta(x - Vt)P \tag{1}$$

Figure 1. Verberie viaduct.

where P is the force running over the bridge at the speed V (δ is the symbol for the Dirac function).

In the case of a mass running over the bridge, the force is replaced by $(m \cdot g)$ and the mass is associated to the term of mass in the first term of the equation.

2 LIMITS CONCERNING THE DYNAMIC COEFFICIENTS

2.1 *Resonance effects observed by SNCF*

On the first high speed line built in France, (see Fig. 1) track deformations appeared above some bridges few weeks after the beginning of the commercial traffic. The speed of the trains was reduced. The bridges were too smooth but their stiffness was within the limits that define the validity domain of the UIC dynamic coefficient.

The vertical acceleration of the decks was greater than 0.7 g further shaking the ballast stones like chestnuts in a frying pan! it resulted in great difficulties to maintain a good geometry of the track.

Some experimentations made by SNCF and later, by the UIC confirmed this 0.7 to 0.8 g limit value. Over this value, the compacity of the ballast is not sufficient and the sleepers cannot resist against the actions (forces) coming from temperature or wheels and vehicles.

Why such phenomena happened? The following chapters may allow answering this question and explaining the knowledge that has been produced and put into the Eurocodes or into the Interoperability Technical Specifications (ITS). A large part of this knowledge concerns the excitation of the structure caused by the trains.

In fact, few studies had focalized on the effect of the entering and coming out of numerous vehicle axles on the bridge and on the resulting vibrations. The excitation function in the partial differential equation has to take into account the characteristic of periodicity due to the succession of the axles. In this case it becomes:

$$P(x,t) = \delta(x - Vt) \cdot (P + Q \cdot \sin(\omega t)) \qquad (2)$$

where ω represents the frequency of the regularly spaced moving axles.

This force $P(x,t)$ may lead to resonance into the bridge, further amplifying deformations with a factor of 50 or 100 (because damping is very low), instead of a maximum value of 2 that is given by the previous classical dynamic coefficient.

3 ANALYSIS OF THIS PHENOMENON

3.1 *General*

There are now many methods to solve the partial differential equation among which the finite elements method and computer programs. Many papers deal with this topic. Calculations are

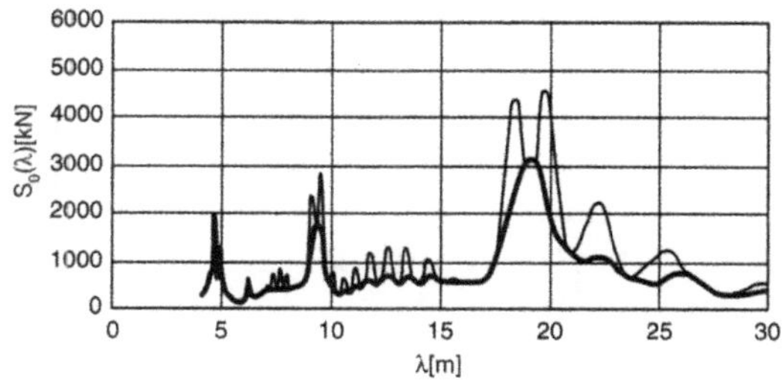

Figure 2. Train signature.

generally made through a modal analysis and the response is obtained by the method of modal superposition. However, only few papers concentrate on the excitation function due to trains running over bridges. Yet, many pieces of information can be found when studying this function. For instance, it allows understanding what was experienced and the resonance phenomena in railway bridges. It also permits to define special "critical" speeds different from the critical speed Vc usually defined in the dynamics handbooks as follow:

$$Vc = 2 \cdot n_0 \cdot L \tag{3}$$

where n_0 is the first natural frequency of vibration of the beam, L its span.

3.2 *Analysis of the excitation function due to trains*

A train, with a length X_n taken between the first and the last axle, can be represented by a succession of n loads P_i the abscissa of which are denoted x_i (distance between the load P_i and the first axle). For an observer, the excitation function due to these moving loads can be decomposed into Fourier series as follows:

$$P(t) = \frac{2}{X_n} \int S_0(\lambda) \left(\cos \frac{2\pi Vt}{\lambda} + \sin \frac{2\pi Vt}{\lambda} \right) d\lambda \tag{4}$$

with λ is the wavelength of the excitation and:

$$S_0(\lambda) = \left[\left(\sum_{k=1}^{n} P_k \cos \frac{2\pi x_k}{\lambda} \right)^2 + \left(\sum_{k=1}^{n} P_k \sin \frac{2\pi x_k}{\lambda} \right)^2 \right]^{1/2} \tag{5}$$

where $S_0(\lambda)$ represents the train excitation and can be considered as a dynamic signing of the train ("signature") (see Fig. 2).

3.3 *Solution to the partial differential equation*

With this representation of the excitation function, simplifications can be made for simply supported girders. It then becomes easy to calculate the parameter of interest, i.e. the maximum vertical acceleration at mid-span when trains are running over the bridge with the speed V. These simplifications can be, for example, to consider only one mode (the first bending one) and the resonance situations. Some similar simplifications are also possible if the other modes (bending or torsion) are important.

If the term of the Fourier series corresponding to resonance is kept as follows:

$$\omega = \frac{2\pi V}{\lambda} \tag{6}$$

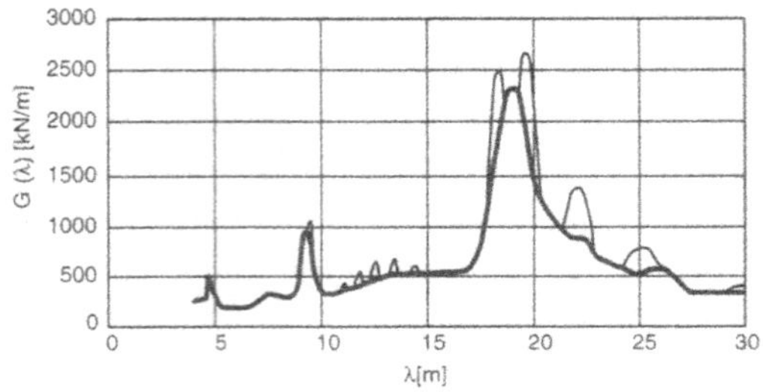

Figure 3. Train spectrum.

the maximum value of the acceleration Γ is obtained by multiplying the following 3 terms:

$$\Gamma = C \cdot A(L/\lambda) \cdot F(\lambda) \tag{7}$$

where C is a constant term depending upon the mass m of the deck:

$$C = \frac{4}{m\pi} \tag{8}$$

$A(L/\lambda)$ depends only on the span L of the deck and is a dynamic influence line defined as:

$$A(L/\lambda) = \left| \frac{\cos \frac{\pi L}{\lambda}}{\left(\frac{2L}{\lambda}\right)^2 - 1} \right| \tag{9}$$

$G(\lambda)$ depends on the so-called "signature" of the train and on the damping (ξ) of the structure. It is the spectrum of the train and the units are expressed in kN/m. (see Fig. 3).

Knowing this spectrum, it is very simple to determine the mid-span acceleration for simply supported girders where speeds becomes critical and resonance occurs. That is always the case because of the numerous modes of the bridges and of the numerous excitation frequencies due to the train. It is also necessary to check that the effect of quick loading modelled by the dynamic coefficient is not greater than these resonance effects.

$$G(\lambda) = Max\left[\frac{1}{\xi x_i} S_{0,i}(\lambda)\left(1 - \exp\left(-2\pi\xi\frac{x_i}{\lambda}\right)\right)\right] \tag{10}$$

with $i = 1$ to n
$S_{0,i}$ (λ) is the "signature" of smaller trains with only i axles.

Many calculations with other methods (computer programs) and many experimentations showed the validity of these formulae.

4 HOW TO USE THE SIGNATURE METHOD

It is possible to determine the critical speeds corresponding to the peak values of the spectrum or the peak values of the "signature" (abscissa $= \lambda_c$). These critical speeds V_c correspond to the maximum values of deformations or accelerations and are defined as follows:

$$V_c = f_0 \lambda_c \tag{11}$$

The maximum value of the acceleration is obtained using Expression (7):

$$\Gamma = C \cdot A(L/\lambda) \cdot F(\lambda_c) \tag{12}$$

The main parameters acting on the resonance phenomena are now identified: type of train, speed, modal characteristics of the bridge... The designer of the bridge can, with these tools, easily adapt the design of his project to satisfy the design criteria (deflection, acceleration).

Another use of this method is the determination, in the background preparation of the EN 1991-2 of the High Speed Load Model (HSLM) also called "Universal train", that is a dynamic equivalent of the static UIC 71 load model.

The HSLM should allow limiting the number of dynamic analyses to proceed if many different high speed trains have to run over the bridges. Indeed, the previous paragraphs show that, theoretically, there would be as many calculations as types of trains and speeds because the response of the bridge depends on the geometrical spacing of the axles-loads, that can infinitely vary.

5 THE HIGH SPEED LOAD MODEL (HSLM)

The definition of the HSLM is based upon some established facts:

- The possibility to get simple "signatures" for regularly spaced axles trains, articulated trains and classical trains (engines + car-bodies);
- The fact that the HSLM should have a signature covering all signatures of existing or future high speed trains. It is well known that many different high speed trains will be built in the near future and, with the new interoperability technical specifications that these trains will run everywhere on the European high speed lines.

If we look now at the envelop curve of the "signature" curves of all possible existing or future trains, it can be seen that there are two separate zones:

- The first one corresponds to the wavelength smaller than 5 m with a very important noise;
- The second one, with a wavelength over 5 m, is a zone where peaks can be distinguished.

These two zones lead to definition of two load models:

- The first one (B type) is a succession of loads regularly spaced with varying loads, varying number of axles and varying spacing in function of the span of the bridge, where span is smaller than 7 m. 7 m has been chosen instead of 5 m in order to limit the number of calculations to make;
- The second one is an articulated train (A type) (see Fig. 4) with varying axle loads, spacing between axles and length of vehicles. It should be used for bridges with span over 7 m. Only 10 load models cover all existing and future trains.

But keeping only ten trains forces to shift slightly the wavelength domain to check, and thus, to shift slightly the speed domain. The step between the ten trains and uncertainties on the natural frequencies of the bridges generates a 20% increase in the maximum speed to be taken into account.

The previous "signature" or the spectrum of excitation $G(\lambda)$ being independent of the structure of the bridge, the afore-mentioned high speed load model can be used for any structural configuration (isostactic beams, continuous beams, arches and so).

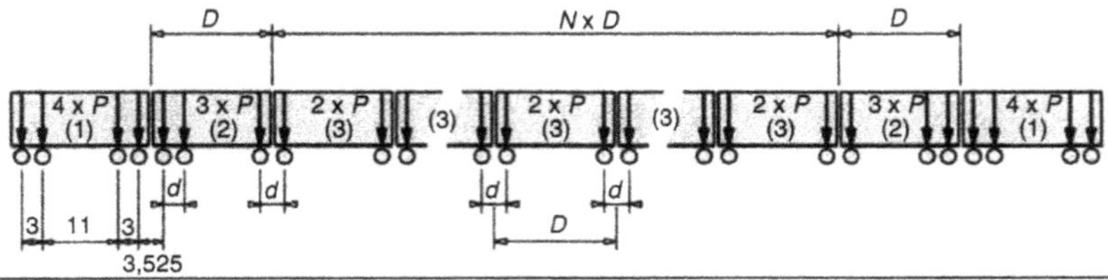

Representative train	Number intermediate cars N	Length of cars D [m]	Distance between axle of bogie d [m]	Force p [kN]
A1	18	18	2.0	170
A2	17	19	3.5	200
A3	16	20	2.0	180
A4	15	21	3.0	190
A5	14	22	2.0	170
A6	13	23	2.0	180
A7	13	24	2.0	190
A8	12	25	2.5	190
A9	11	26	2.0	210
A10	11	27	2.0	210

Figure 4. Train HSLM – A.

6 SIMPLY SUPPORTED BRIDGES

It is possible for the simply supported bridges to limit the calculations when choosing among the ten load models the one that will create the maximum response of the bridge (deflections, acceleration).

A simplified formula for the mid-span acceleration was previously given in (7). The product:

$$A\left(\frac{L}{\lambda}\right) \cdot G(\lambda) \tag{13}$$

depends only on the span of the bridge and on its damping. It does not depend on other bridge mechanical or geometrical parameter.

So, for every couple of values (L, ξ), it is possible to draw a curve so called "agressivity" curve for the ten type. A load models, and find out which model (1 to 10) create the maximum response for each value of λ.

$$Max\left[A\left(\frac{L}{\lambda}\right) \cdot G(\lambda)\right] \tag{14}$$

These curves have been drawn by the Structural Experts of the UIC.

Given a bridge (L, ξ, f_0) and the maximum speed V_M of the trains, it is possible to determine on the adequate (L, ξ) curve where:

- The authorised domain

$$\lambda \leq \frac{V_M}{f_0} \tag{15}$$

- Inside this domain, the maximum value of the "agressivity" curve and the corresponding wavelength (λ_M) and thus, the load model that will give the maximum acceleration (see Fig. 5).

Then, the maximum acceleration at mid-span can be calculated with this type A load model at the speed $V = \lambda_M \cdot f_0$.

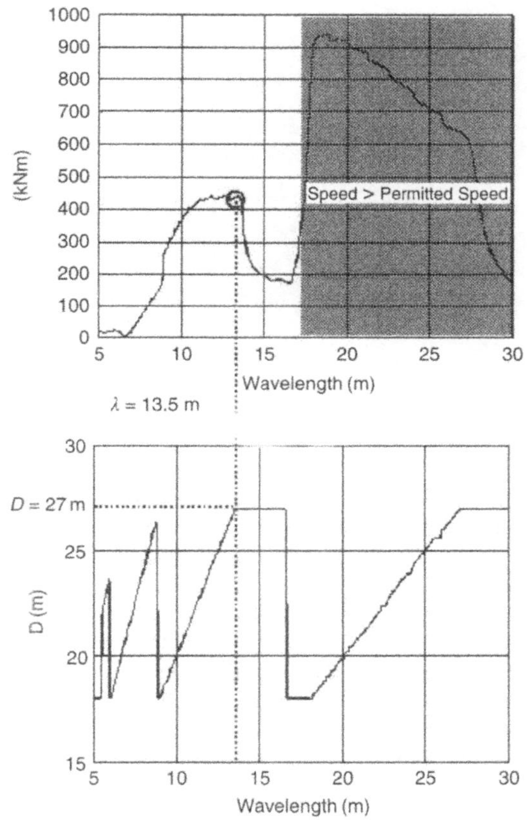

Figure 5. Agressivity curve.

7 CONCLUSIONS

Dynamic analyses indicate clearly two important elements:

- The fundamental importance of Service Limit State criteria for designing rail-way bridges. It is obvious that checking resistance and durability through Ultimate Limit States or Fatigue Limit States is not sufficient to guarantee that specified service requirements are satisfied. These special requirements are strongly needed. They must be completed by the checking of other criteria like comfort criteria or track stability or track-bridge interaction criteria (i.e. the consequences on track stability or on rail stresses of the longitudinal displacement of the bridge).
  Concerning track stability, it is important to remark that it is some sort of ultimate limit state for track (and safety problems for vehicles running over it!) while it concerns Service Limit States for the bridge. This the real meaning of theses limit states.
- The "art" of the engineer is not only to use more and more sophisticated finite elements computer programs. These numerical methods allow him checking, making more accurate, and validating the intuition or the opinion he can build concerning his design problems. But also, except if he makes a lot of calculations and if he is able to analyse correctly the whole package of results and find the influence of the determining parameters, these tools do not often allow him to directly understand the physical phenomena.

Whatsoever, these computer programs are based upon well known physics and mathematical equations. The study of these equations also allows accessing the understanding of the phenomena.

Figure 6. Arc viaduct.

But it is necessary to know how to use the mathematical tools permitting to solve these equations or to find simplified methods leading to the results. Many experiments by trial and error processes are needed for this. The success is obtained where the fundamental equations are not forgotten. Computer calculations are only a complementary method that gives, of course, more accurate results but cannot be a self objective for the process.

The dynamic behaviour of the railway bridges could not be understood totally without finalizing the previous studies that are based upon a simple understanding of the physical behaviour of the bridges using the solving of the equations with mathematical tools and adequate simplifications where possible. The Engineer will produce a good design only if he understands completely the physics of the behaviour of the railway bridges (see Fig. 6).

REFERENCES

EN1991-2 (2003) – "Actions on Structures – Part 2: General Actions – Traffic loads on bridges", European Committee for Standardization (CEN), Brussels.

ERRI D214/RP9 (2001) – "Railway bridges for speeds >200 km/h", European Rail Research Institute (ERRI), Final Report, Utrecht.

CHAPTER 2

Design issues related to dynamic effects for high speed railway bridges in Spain

J.M. Goicolea-Ruigómez & F. Gabaldón Castillo
Escuela de Ingenieros de Caminos, Technical University of Madrid, Spain

ABSTRACT: High-speed railway lines are an important factor for modernisation of transport in Europe, specially in countries such as Spain with a historic deficit in infrastructure. This work will also discuss some issues related to dynamic effects in railway bridges. Firstly we review and comment analysis methods from a practical point of view. Following, issues related to train types are discussed. Finally, some considerations related to the structural type of the bridge are studied.

1 INTRODUCTION

High speed railway lines are currently the main element in new transport infrastructure, contributing to improved opportunities for travel and commerce in Europe. They already provide an efficient transport link between some of the main Spanish cities, and will soon link Spain and Portugal together and with the rest of Europe.

One of the main design issues associated specifically to the design of bridges and structures in railway lines are the dynamic effects due to moving loads from train traffic. This has been studied since the early days of railways [1], as well as in more recent works (e.g. Fryba [2] and Alarcón [3]).

The design codes existing up to now [5,6,7] for design of railway bridges consider the dynamic response through an impact factor, which represents the increase in the dynamic response with respect to the static one for a single moving load.

However, high speed railway lines pose dynamic problems of higher order, due to the possibility of resonance from high speed traffic at speeds above 200 km/h, considering the typical distances between axles in railway coaches and the main eigenfrequencies of bridges. Resonance occurs when the excitation frequency coincides with that of the fundamental vibration mode of the bridge. This may be quantified through the so called *wavelength* of excitation, $\lambda = v/f_0$, where f_0 is the first natural frequency of deck vibration and v the train speed. Resonance occurs when the characteristic length D_k of separation between axles coincides with a multiple of the said wavelength:

$$\lambda = \frac{D_k}{i}, \quad i = 1,2,3,4. \quad \Rightarrow \quad \text{resonance} \tag{1}$$

Within Europe a joint effort for research and study of dynamic effects in high speed lines has been carried out within the European Railway Research institute (ERRI) by subcommittee D214 [13]. These and other findings have been included in the recently drafted engineering codes [8], [10] and [9].

In this paper we consider some issues related to our experience in Spain for the dynamic effects on railway bridges. Firstly (section 2) we discuss the various analysis methods for calculation and discuss the practical issues of their application. Following (section 3) we consider issues related to train types influencing the dynamic response. Finally (section 4), some considerations are made related to structural type of the bridge.

2 ISSUES RELATED TO ANALYSIS METHODS

2.1 *Impact factor* Φ

The basic method followed up to now in the existing engineering codes for railway bridges [5,6,7] has been that of the impact factor, generally represented as Φ. As has been discussed previously in section 1, such coefficient represents the dynamic effect of (single) moving loads, but does not include resonant dynamic effects.

The dynamic increment φ' for a single moving load at speed v on an ideal bridge (i.e. without track or wheel irregularities) is evaluated in [7] by the following expression:

$$\varphi' = \frac{K}{1-K+K^4}; \quad K = \frac{\lambda}{2L_\Phi} \tag{2}$$

where L_Φ is the equivalent span of the element under study and $\lambda = v/f_0$ the wavelength of excitation. According to Equation (2) the value of the dynamic increment reaches a maximum value of $\varphi'_{\max} = 1.32$, for $K = 1.76$. The final impact factor takes into account additionally the effect of irregularities through an additional term (φ''):

$$\Phi \geq \max\left(1 + \varphi' + \varphi''\right) \tag{3}$$

The impact factor so defined is applied to the effects obtained for the static calculation with the nominal train type (LM71):

$$\Phi \cdot E_{\text{sta,LM71}} \geq E_{\text{dyn,real}} \tag{4}$$

We remark that the impact factor Φ is applied not to the real trains, but to the effects of the LM71 load model, which is meant as an envelope of passenger, freight traffic and other special trains, generally much heavier than passenger trains. Finally, the applicability of impact factor Φ is subject to some restrictions, involving bounds for f_0 as well as a maximum train speed of 200 km/h [6].

2.2 *Simplified models based on dynamic train signature*

The so-called *dynamic train signature* models develop the response as a combination of harmonic series, and establish an upper bound of this sum, avoiding a direct dynamic analysis by time integration. In counterpart their application is limited to *simply supported bridges*, which can be represented dynamically by means of a single harmonic vibration mode. Their basic description may be found in [13].

All these methods furnish an analytical evaluation of an upper bound for the dynamic response of a given bridge, as a product of three terms: a constant term, a *dynamic influence line* of the bridge, and a *dynamic signature* of the train. Let us take as an example the LIR method [13] for evaluating the maximum acceleration. This procedure is based on the analysis of the residual free vibrations after each individual single load crosses a simply supported bridge. The acceleration Γ at the centre of the span is given by:

$$\Gamma = C_{\text{acel}} \cdot A(K) \cdot G(\lambda) \tag{5}$$

$$A(K) = \frac{K}{1-K^2}\sqrt{e^{-2\zeta\frac{\pi}{K}} + 1 + 2\cos\left(\frac{\pi}{K}\right)e^{-\zeta\frac{\pi}{K}}} \tag{6}$$

$$G(\lambda) = \max_{i=1}^{N}\sqrt{\left[\sum_{x_1}^{x_i} F_i \cos(2\pi\delta_i)e^{-2\pi\zeta\delta_i}\right]^2 + \left[\sum_{x_1}^{x_i} F_i \sin(2\pi\delta_i)e^{-2\pi\zeta\delta_i}\right]^2} \tag{7}$$

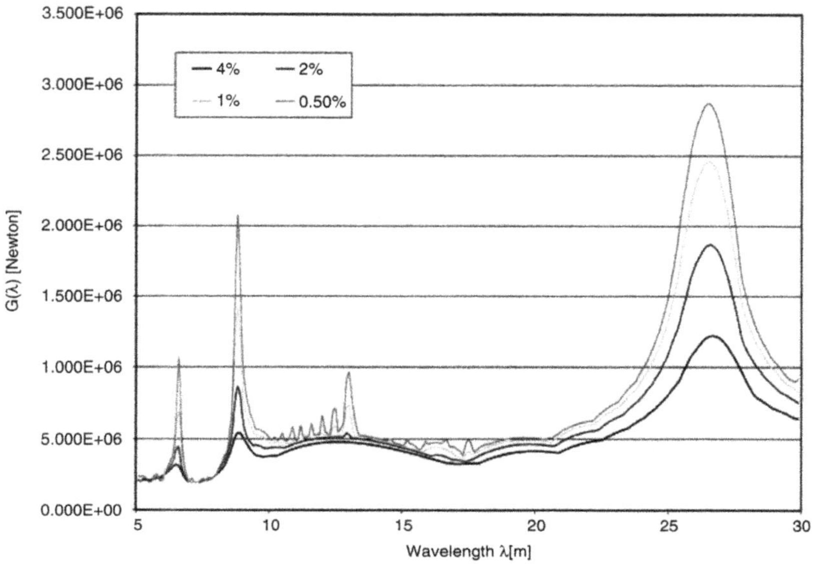

Figure 1. Dynamic signature (LIR) of ICE2 train for different damping rates. The peak response is attained for excitation wavelengths near the coach length.

In these expressions $C_{\text{acel}} = 1/M$ is a constant (the inverse of the total mass of the bridge), ζ is the damping rate, x_i are the distances of each one of the N load axes F_i to the first axis of the train, and $\delta_i = (x_i - \underline{x}_1)/\lambda$.

The term $G(\lambda)$ (Eq. (7)) is the *dynamic signature* referred to above. It depends only on the distribution of the train axles and the damping rates. Each train has its own dynamic signature, which is independent of the mechanical characteristics of the bridges. As an example, Figure 1 represents the dynamic signature of train ICE2, for different values of damping.

The term $A(K)$ (Eq. (6)) defines a function of K (itself dependent on speed v), called the bridge *dynamic influence line*. It depends solely on the span L, the first natural frequency (f_0) and damping (ζ).

Neither C_{acel} nor $A(K)$ depend on the characteristics of the train. Separating the contributions from the bridge and those from the train ($G(\lambda)$, dynamic signature), it is possible to determine easily the critical parameters of span and wavelength for which the dynamic response of the bridge is maximum.

As has been said before, these dynamic signature methods have been developed in principle for simply supported, isostatic bridges. However, some studies have been carried out which in some cases extend their applicability to certain classes of redundant structures. For instance, Liberatore [15] has developed dynamic signature methods to establish the *modal aggressivity* of continuous decks with 2 spans.

2.3 *Dynamic analysis with moving loads*

This general class of models are based on time integration of the dynamic equations for the structure, subject to a series of moving loads of constant values, representative of each axle of a given train. The structure is first discretised using a finite element method or a more particular bar technique, arriving to a vector d of N (unknown) nodal displacements. The model for the structure may be analysed either through an direct integration of the complete system,

$$\boldsymbol{M} \cdot \ddot{\boldsymbol{d}} + \boldsymbol{C} \cdot \dot{\boldsymbol{d}} + \boldsymbol{K} \cdot \boldsymbol{d} = \boldsymbol{f}(t), \tag{8}$$

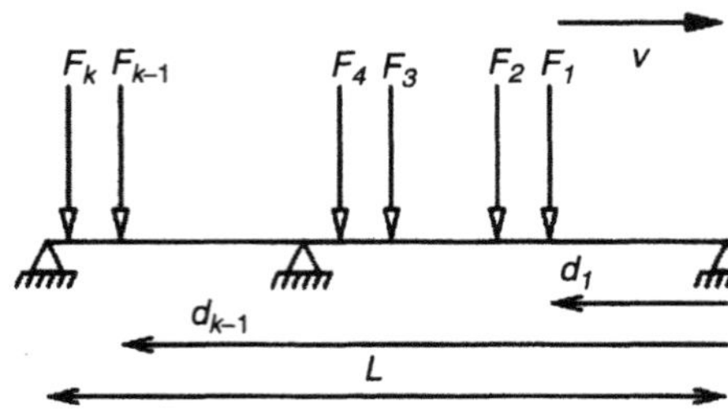

Figure 2. Load model for a train of moving loads.

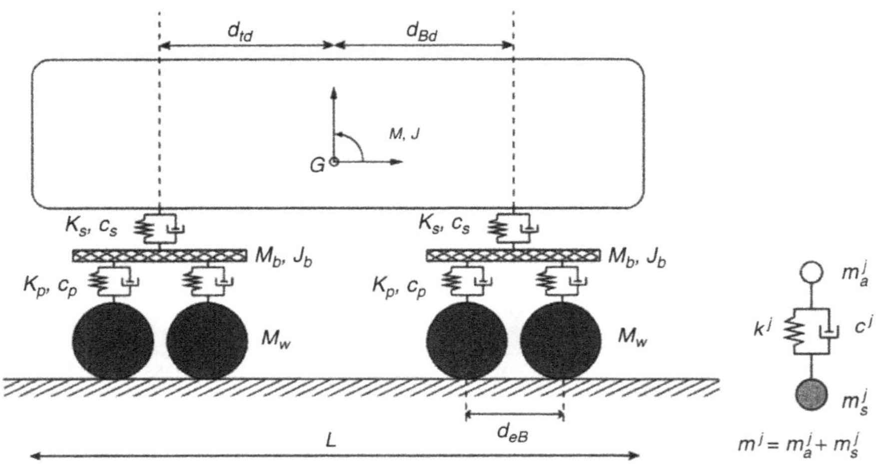

Figure 3. Vehicle–structure interaction models: (a) full interaction model; (b) simplified interaction model.

where $\boldsymbol{M}$ is the mass matrix, $\boldsymbol{C}$ the damping matrix, $\boldsymbol{K}$ the stiffness matrix, and $\boldsymbol{f}(t)$ the external load vector. In the case of moving loads, this load vector must represent the load histories of each of the train axles (F_k, $k = 1, \ldots, n_{ax}$) at the particular speed considered (Fig. 2).

A reduction of degrees of freedom may be performed through a modal analysis. Modal reduction reduces substantially the number of equations to integrate, and may be performed through an approximate numerical procedure to obtain the eigenmodes of vibration. This capability is provided by most commercial finite element programs. Alternatively the modal reduction may be achieved through an analytical (closed form) calculation, for certain cases of simple structures.

In references [4, 11, 16, 18] more details are given regarding the numerical algorithms.

Except for particular cases of simple structures the above Equation (8) must be evaluated numerically by finite element methods. These provide an efficient method for calculation in arbitrary structures. Adequate procedures for pre-processing (definition of load histories for all individual axles) and post-processing are necessary for their practical use in engineering design work [18].

2.4 *Dynamic analysis with vehicle-structure interaction*

The consideration of the vertical motion of the vehicles with respect to the bridge deck allows for a more realistic representation of the dynamic overall behaviour. The train is no longer represented by moving loads of fixed value, but rather by point masses, bodies and springs which represent wheels, bogies and coaches. In some cases these models may have a non negligible influence on the dynamic response of the bridge.

A general model for a conventional coach on two bogies is shown in Figure 3a, including the stiffness and damping (K_p, c_p) of the primary suspension of each axle, the secondary suspension

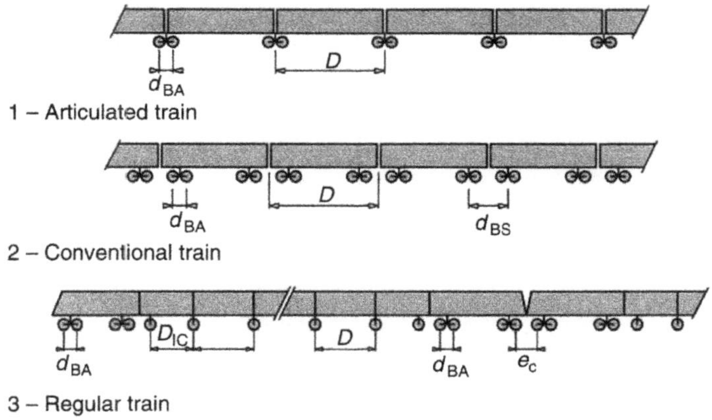

Figure 4. Different types of high-speed trains, according to Eurocode 1 [10].

of bogies (K_s, c_s), the unsprung mass of wheels (M_w), the bogies (M_b, J_b), and the vehicle body (M, J). Similar models may be constructed for articulated or regular trains.

The level of detail in the above model is not always necessary. Often simplified models may be employed in which for each axle only the primary suspension, equivalent unsprung and sprung masses are considered (Fig. 3b). In this model each axis is independent from the rest, thus neglecting the coupling provided by the bogies and vehicle box, as well as the rocking motion of the vehicle box. Further details of these models are described in [11].

The consideration of vehicle structure interaction often reduces the dynamic response of the bridge, as is shown in [4,16] (see also Fig. 9). This may be explained considering that part of the energy from the vibration will be transmitted from the bridge to the vehicles. In short span bridges (i.e. $L \leq 30$ m) this may produce a reduction of up to 30% [4]. However, in longer spans or in continuous deck bridges the reduction will generally be small. As a consequence it is not generally necessary to perform dynamic analysis with interaction, except for particular cases.

3 ISSUES RELATED TO TRAINS

As has been mentioned, the configuration of trains and axle spacing is one of the key factors defining the dynamic effects and the possibility of resonant response in bridges. With regard to distribution of axles and bogies three types may be identified in European high speed trains (Fig. 4):

1. Articulated trains: each two coaches share one bogie between them. This type includes current European models Thalys, AVE and Eurostar.
2. Conventional trains: each coach has two bogies. This includes Ice2, Ice3, Etr-y, Virgin.
3. Regular trains: coaches are supported not on bogies but on single axles in the junction between each two of them. This is the case of TALGO.

To ensure dynamic performance not only for the above trains but also for their possible variations and future developments through dynamic analysis ("brute force method") would be extremely costly as well as of doubtful efficiency. Small variations in the configuration of a given train may influence significantly the resonant peaks, making it extremely difficult to assure the fulfilment of the dynamic performance interoperability conditions.

The concept of *train signature* (Section 2.2) is very useful for the purpose of understanding the dynamic effects of each particular train, and for comparing the responses. This train signature is a characteristic of each particular train, independent of the bridge. Figure 6 shows the dynamic signature (DER method) obtained for the most common current European high-speed

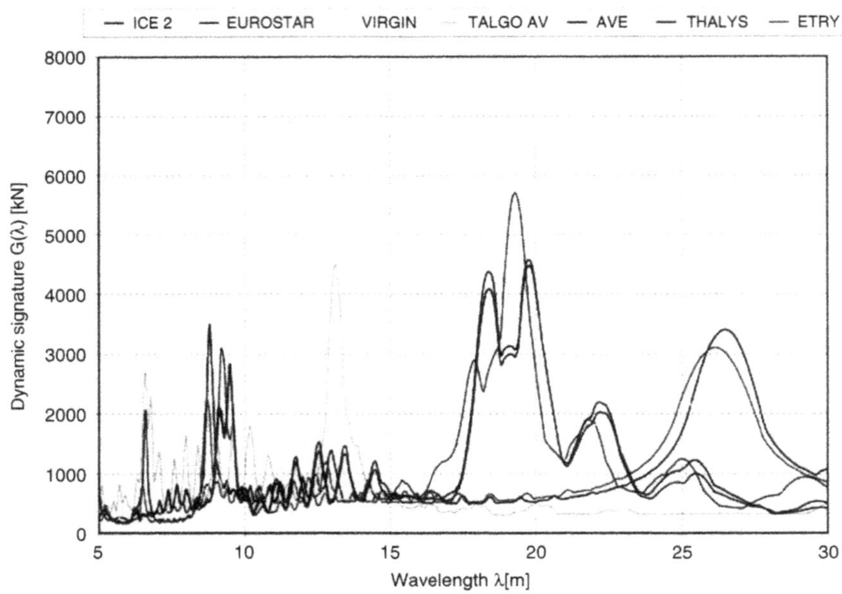

Figure 5. Dynamic signatures (zero damping) for European high-speed trains, including articulated trains (EUROSTAR, AVE, THALYS), conventional trains (ICE2, VIRGIN, ETRY) and regular trains (TALGO).

trains. The dynamic response of the bridge may then be obtained by multiplying the train signature by the so-called *dynamic influence line* of the bridge. In Figure 5 one can clearly see the variation in the response for the different classes of trains. The articulated trains show response peaks for wavelengths between 17 and 21 m. The conventional trains exhibit the highest response for wavelengths between 24 and 28 m. Finally, the regular train TALGO shows the peak for 13 m. These critical wavelengths are related in each case to the length of the wagons defining characteristic axle or bogie spacings. In effect, the shortest coach length is that of the TALGO, with axle spacings of 13.14 m.

An envelope of these signatures may be easily obtained, as shown in Figure 6. This envelope may be used for design purposes, in order to determine the maximum response of a given bridge in which any of the above high speed trains may circulate. However, this kind of design procedure has two main drawbacks:

1. It is only valid for the cases in which the train signature is applicable, that is simply supported isostatic bridges.
2. It covers the current catalogue of high-speed trains, but does not include possible variations which may be incorporated in the future by the constructors.

It is highly desirable from a social and economical point of view that the high-speed line infrastructure is interoperable, that is all high speed trains from other European lines may also use them even though the line was not foreseen initially specifically for them. From the point of view of structural requirements on bridges the static strength is assured by the static load model LM71. The dynamic performance must be assured by a set of dynamic analyses that covers all possible present (and future) trains.

The task of developing a High Speed Load Model (HSLM) which would ensure interoperability conditions was performed by ERRI D214.2 [14], which first drafted envelopes of DER signatures for all current high-speed trains and their possible variations.

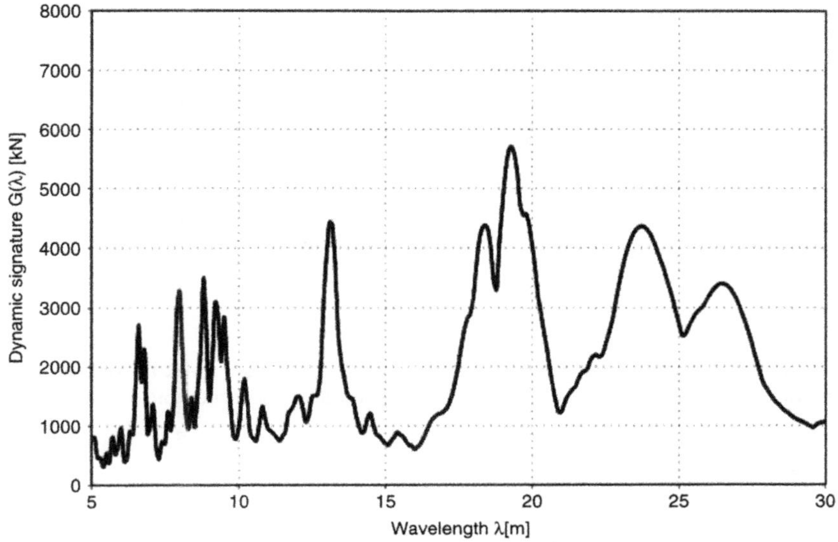

Figure 6. Envelope of dynamic signatures (zero damping) for European high-speed trains.

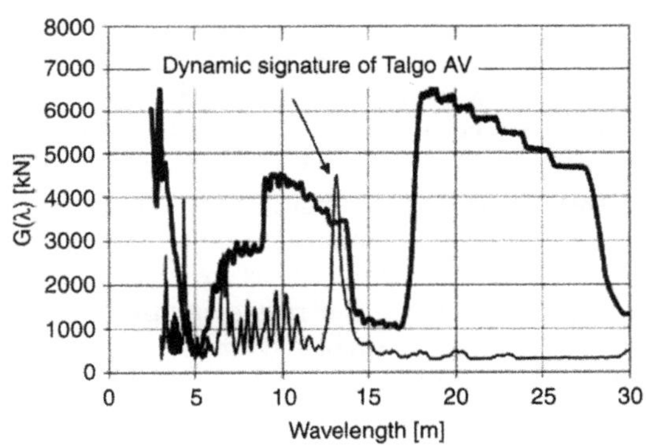

Figure 7. Dynamic signatures of the initial UNIV-A envelope and of Talgo AV.

As an example, we document the fit of TALGO HS train within the initially considered UNIV A envelope (Fig. 7), which included only articulated and conventional trains. It may be seen that for short wavelengths ($\lambda < 15$) it does not cover the TALGO signature. In order to cover also regular trains, the final version of this envelope and the resulting HSLM-A family of trains was modified accordingly.

Following, a family of fictitious articulated trains (*Universal trains*) was devised ensuring that their signature envelope effectively covered the signatures of all real trains. This has lead to the HSLM-A and HSLM-B families of universal trains [4,10].

4 ISSUES RELATED TO BRIDGES

Some brief aspects regarding the main differences in behaviour to be expected from different structural types of bridges are discussed here.

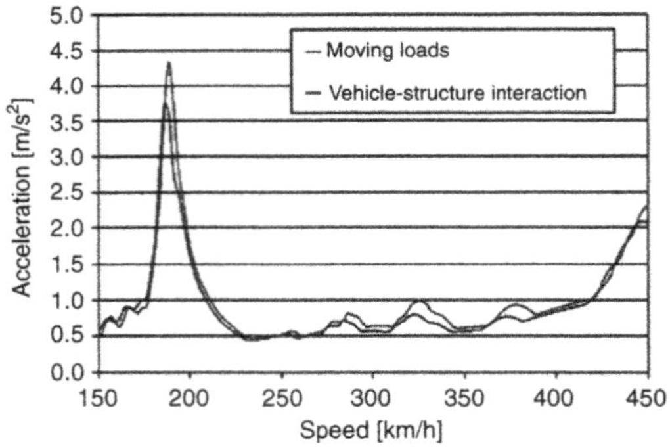

Figure 8. Maximum acceleration at centre of span as a function of train speed for TALGO AV. $L = 20$ m, $m = 7$ t/m, $EI = 453919$ kN/m^2, $f_0 = 16$ Hz, $\zeta = 1\%$. Results with and without vehicle-bridge interaction.

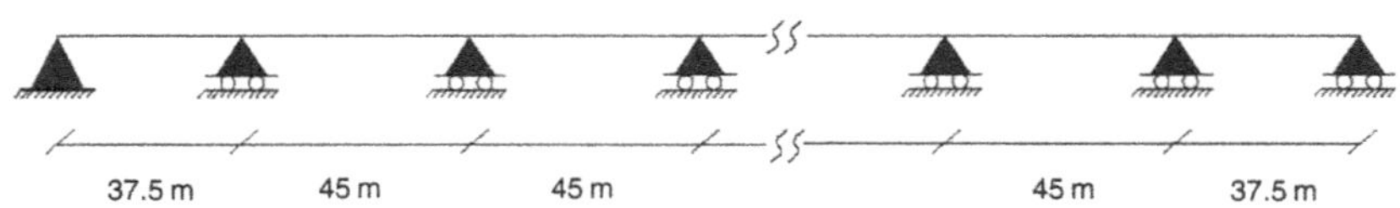

Figure 9. Geometric characteristics of continuous deck viaduct.

We first consider the dynamic response of a simply-supported isostatic bridge (Fig. 8). A very sharp and intense resonant peak in the response is obtained for a speed around 200 km/h, for which the main bridge frequency equals that of the travelling loads. The dynamic response of isostatic simply-supported bridges is determined mainly by the first mode of vibration, higher modes have a very small participation in the response. However, we remark that for the purpose of calculating reaction forces and section moments higher modes have a larger influence [16]. Further, a remarkable feature of the response is that the effects are maximum at a critical train speed, and do not increase for larger speeds. In fact, it is clearly seen that the response at even 400 km/h is considerably lower to that at the critical speed of 200 km/h.

In current high-speed lines it is very common to have fairly long viaducts of between 100 to 2000 m. Continuous deck bridges are built very often in these cases, as they have a number of advantages. One of the main benefits is from the construction point of view, which very often is by pushing the deck being fabricated at one of the extremes. An important consideration to be made is that of the deck-track interaction, in order to decide the placement of fixed neutral points and track dilatation devices. Finally, from a dynamic point of view, continuous deck bridges exhibit a much different (and generally lower) response to traffic loads than isostatic bridges. This is due to the fact that at any time the response of the bridge is governed not by one vibration mode but by a large number of modes. These modes are excited positively or negatively by traffic loads at the various spans, and the sum of the responses cancels much of their effects.

As an example we present some results for a viaduct in the Madrid-Barcelona line. It is a continuous beam with 9 spans, the outer ones of 37.5 m span and the central ones of 45 m, a total length of 390 m, with a slight curve in plan view (Fig. 9).

For this viaduct a complete dynamic modal analysis was carried out under the actions of high speed trains in Europe. In Figure 10 we show the results for the AVE Spanish HS train. In this figure we also show the individual contributions from the individual modes of vibration.

Two main consequences may be drawn from these results. Firstly, in broad terms the maximum structural response does increase monotonically with the higher speeds, being larger for the highest speeds. However, in this case significant modal resonance is obtained for modes 3 and 5 which distort this character slightly.

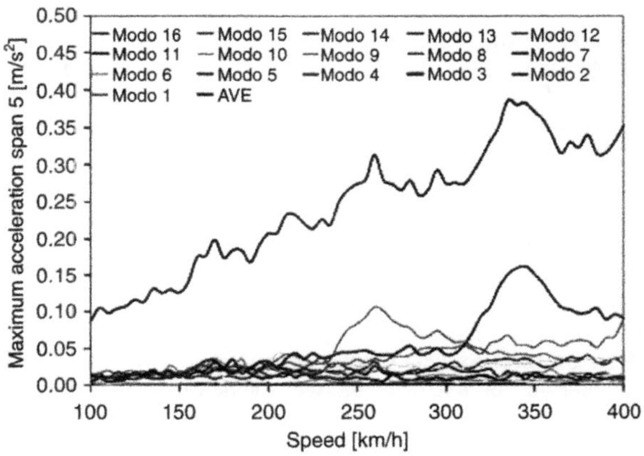

Figure 10. Maximum displacement at the center of span 5 versus speed for a double composition AVE HS train (black line). Also shown in colours are the contributions of the various modes with $f \leq 20$ Hz.

Secondly, the peak structural response is much lower to that which would be obtained by isostatic simply supported independent spans.

In this viaduct it was performed a complete modal analysis of the structure under the action of six of the most important European high speed trains (Fig. 10).

5 CONCLUDING REMARKS

As a consequence of the work described above we point out the following remarks:

- Dynamic effects in general and the possibility of resonance in particular require in general a dynamic analysis for the design of high speed railway bridges.
- Simplified models which provide upper bounds for dynamic effects are of limited applicability. Moving load finite element models or even vehicle-structure interaction models for more special cases provide a general methodology.
- The consideration of dynamic vehicle-structure interaction leads to more realistic predictions, in the case where adequate data from the trains are available to build such models. The structural response predicted is somewhat lower to that of moving load models for resonant scenarios. It is these resonant situations that generally limit the design.
- Hyperestatic continuous deck bridges lead generally to a less marked resonance, although a dynamic analysis is still necessary for them. In practice, HSLM models for interoperability of railway lines are adequate bounds of the dynamic effects in the cases studied.
- It is necessary co consider both signs of dynamic effects, including also the dynamic uplift which may be significant in some design scenarios. This may be done through specific design provisions or through a special unloaded train.

REFERENCES

[1] Timoshenko, S.P. & Young, D.H. 1955. *Vibration problems in engineering*. Van Nostrand, NY (3.ª ed.).
[2] Fryba, L. 1996. *Dynamics of railways bridges*. Thomas Telford.
[3] Alarcón, E. 1971. *El coeficiente de impacto en puentes de ferrocarril*. Revista de Obras públicas, September.
[4] Goicolea, J.M. 2004. Dynamic loads in new engineering codes for railway bridges in Europe and Spain. *Workshop on Bridges for High-Speed Railways*, Porto, 3–4 June.

[5] Ministerio de Obras Públicas y Urbanismo. 1975. Instrucción ralativa a las acciones a considerar en el proyecto de puentes de ferrocarril.
[6] Comité Europeo de Normalización. 1998. Eurocódigo 1: Bases de proyecto y acciones en estructuras; parte 3, acciones de tráfico en puentes, traducido y publicado por AENOR.
[7] Union Internationale des Chemins de Fer (UIC). 1979. UIC Code 776-1 R: Charges a prendre en consideration dans le calcul des ponts-rails.
[8] Ferrovie dello Stato. 1997. Sovraccarichi per il calcolo dei ponti ferroviari.
[9] Instrucción de acciones a considerar en el proyecto de puentes de ferrocarril. 2003. Ministerio de Fomento. Borrador, pendiente de publicación.
[10] European Committee for Standardization. 2003. EN 1991-2: EUROCODE 1 - Actions on structures, Part 2: Traffic loads on bridges.
[11] Domínguez J. 2001. *Dinámica de puentes de ferrocarril para alta velocidad: métodos de cálculo y estudio de la resonancia*. Tesis Doctoral. Escuela Técnica Superior de Ingenieros de Caminos, Canales y Puertos de Madrid (UPM). Publicada por la Asociación Nacional de Constructores Independientes (ANCI).
[12] Clough, R. & Penzien, J. 1993. *Dynamics of Structures*. 2nd. edition, Mc Graw-Hill.
[13] Comittee ERRI D214. 1998. Design of Railway Bridges for Speed up to 350 km/h; Dynamic loading effects including resonance; Final report. Draft C. European Rail Research Institute (ERRI).
[14] Comittee ERRI D214.2. 2002. Utilisation de convois universels pour le dimensionement dynamique de ponts-rails. Synthèse des résultats du D214.2 (Rapport final). European Rail Research Institute (ERRI).
[15] Liberatore, D. 2000. Universal train for interoperability analysis; interoperability analysis of continuous decks, ERRI D214 report, Paris May 4.
[16] Goicolea J.M., Domínguez J., Navarro J.A. & Gabaldón F. 2002. *Comportamiento dinámico de puentes de ferrocarril de alta velocidad*. Depto. de Mecánica de Medios Continuos de la Univ. Politécnica de Madrid, Informe técnico para el Ministerio de Fomento.
[17] Goicolea J.M., Navarro J.A., Domínguez J. & Gabaldón F., *Resumen informe Solicitaciones máximas y mínimas*. Depto. de Mecánica de Medios Continuos de la Univ. Politécnica de Madrid, Informe técnico para la comisión de redacción de la IAPF.
[18] Gabaldón F. 2004. *Programa ffcc-load. Manual de usuario*. Grupo de Mecánica Computational, E.T.S. de Ingenieros de Caminos, Univ. Politécnica de Madrid.

CHAPTER 3

Dynamics of bridges under moving loads (past, present and future)

L. Frýba
Institute of Theoretical and Applied Mechanics, Academy of Sciences of the Czech Republic, Prague, Czech Republic

ABSTRACT: The dynamics of bridges has been followed since the construction of the first railway bridges, the effects of steam locomotives up to the sleeper and cross girder effects. The basic dynamic characteristics of bridges (natural frequencies and damping) are represented in figures. The resonant vibration has occurred on bridges under high speed trains in recent time which awake the intensive vibration as well as the necessity to assess the bridges at fatigue. The possibilities of the bridge damping are discussed.

1 INTRODUCTION

The discipline "Dynamics of Bridges" arose during the construction of the first railway bridges in England in the first half of the 19-th century. At that time, the first experimental [1] and theoretical studies [2] were published. Many authors devoted to the problem in the 19-th and 20-s centuries. Among them, Timoshenko [3], Inglis [4] and Koloušek [5] should be mentioned due to their important contributions to the progress.

The steam locomotives excited a harmonic force

$$F(t) = F_0 \sin \Omega t \tag{1}$$

coming from the unbalanced counterweights on the driving wheels. The force (1) was a cause of intensive vibration of railway bridges at steam time.

The deflection-time histories of a steel bridge of the span 56.56 m under a steam locomotive 524.1 are shown in the Figure 1 (experiments) and Figure 2 (theory). The theory corresponded well to the theory, [6]. The dynamic response of a beam to the force (1) is described by the equation (at the resonant condition $\Omega = \omega_1$)

$$v(x,t) = v_0 \frac{F_0 \omega_1}{2F} \frac{\cos \omega_1 t}{\omega^2 + \omega_b^2} [\omega(\cos \omega t - e^{-\omega_b t}) - \omega_b \sin \omega t] \sin \frac{\pi x}{l} \tag{2}$$

2 PAST

The international research organizations ORE (Office for Research and Experiments of the International Union of Railways), ERRI (European Rail Research Institute) and OSZhD (Organization for Railways Cooperation) have solved several problems of the interaction of bridges with moving locomotives and trains after the second world war.

The basic dynamic characteristics of railway bridges as functions of span l coming from that research are summarized in the Figure 3 (f_1 – first natural frequencies), Figure 4 and Figure 5

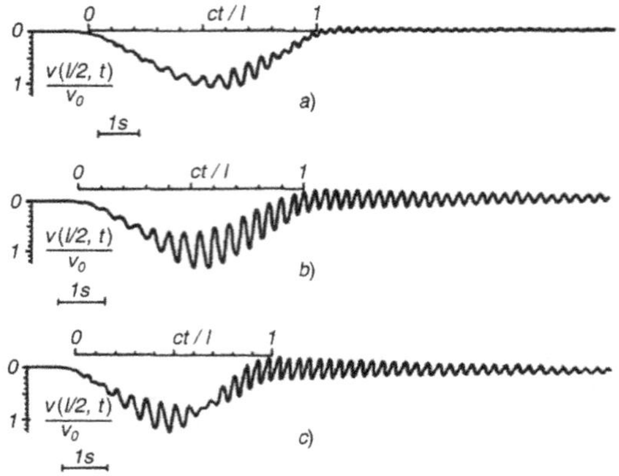

Figure 1. Steel bridge, $l = 56.56$ m, (experiments).

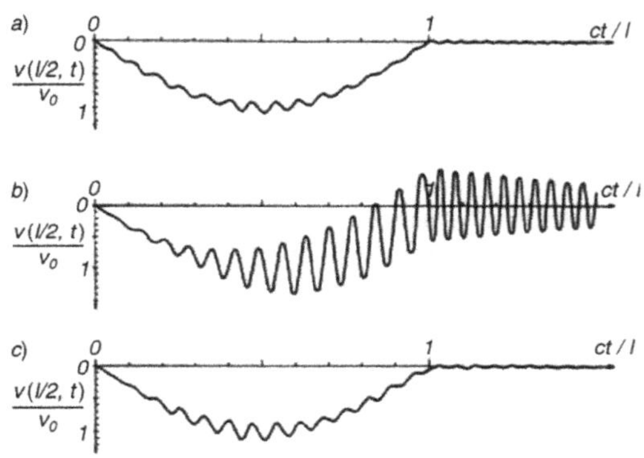

Figure 2. Steel bridge, $l = 56.56$ m (theory).

(ϑ – logarithmic decrements of damping). The functions may be covered by empirical formulas as shown in [7] including the bands with the 95% reliability.

After putting the electric and dieselelectric locomotives with bogies in traffic at speeds under 200 km/h, the cross girder and sleeper effects appeared as the source of bridge vibration because the wheels followed a curved roadway. The Figures 6, 7 and 8 explain the phenomenon.

Therefore, the first theoretical model for the computer was set up (Fig. 9) for an ORE question in sixties, [7], and the dynamic impact factor plotted against the speed (Fig. 10).

The results were compared to the DB (Fig. 11) and SNCF (Fig. 12) experiments. The theoretical and experimental researches are described in details in [7] and in ORE reports.

The following equations describe the dynamic motion of the model in the Figure 9:

$$-I\frac{d^2\overline{\varphi}(t)}{dt^2} + \sum_{i=1}^{2}(-1)^i D_i[Z_i(t) + Z_{bi}(t)] = 0, \tag{3}$$

$$-m_3\frac{d^2 v_3(t)}{dt^2} - \sum_{i=1}^{2}[Z_i(t) + Z_{bi}(t)] = 0 \tag{4}$$

$$P_i + P_{3i} - m_i\frac{d^2 v_i(t)}{dt^2} + Z_i(t) + Z_{bi}(t) - \overline{R}_i(t) = 0; i = 1,\ 2, \tag{5}$$

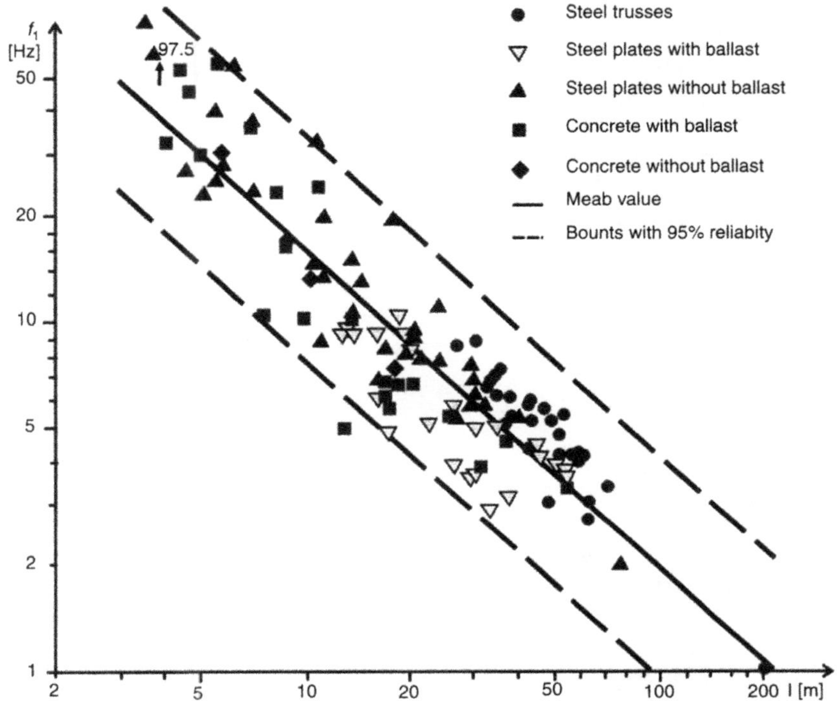

Figure 3. First natural frequencies f_1 as a function of the span l.

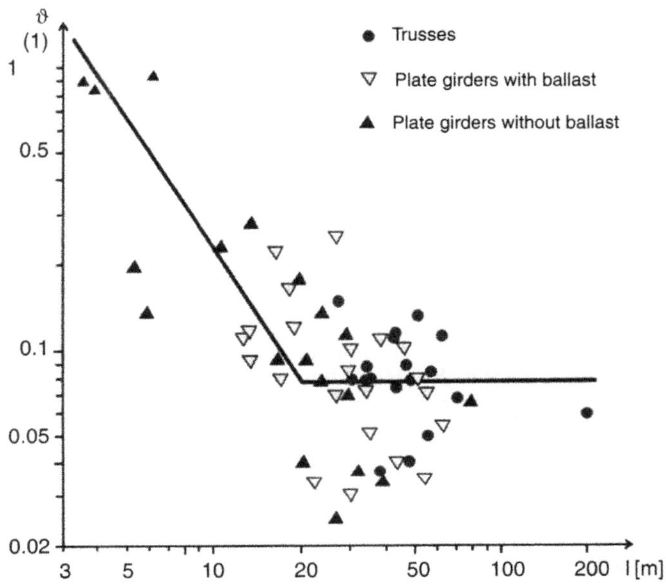

Figure 4. Logarithmic decrements ϑ of damping, steel bridges.

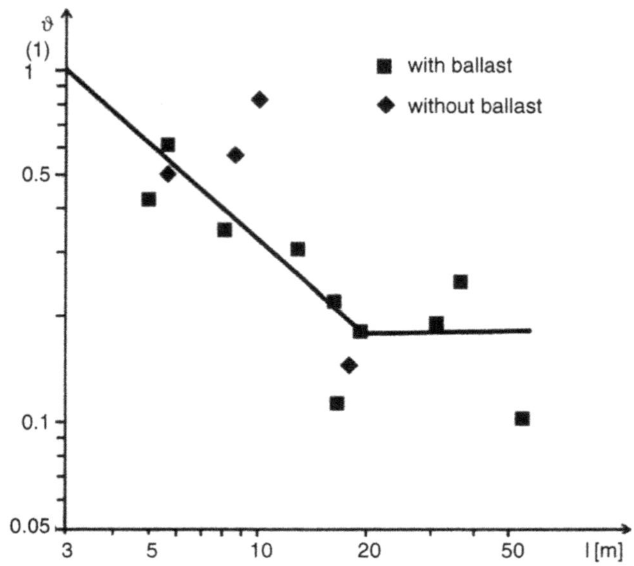

Figure 5. Logarithmic decrements ϑ of damping, concrete bridges.

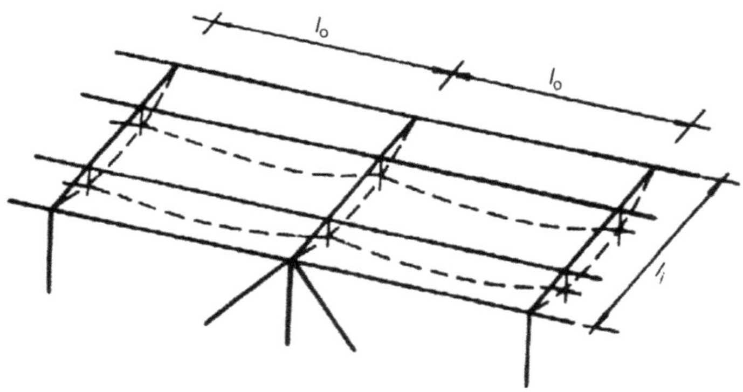

Figure 6. Deflection of the deck grillage, cross beam effect.

$$EJ\frac{\partial^4 v(x,t)}{\partial x^4} + \mu\frac{\partial^2 v(x,t)}{\partial t^2} + 2\mu\omega_b\frac{\partial v(x,t)}{\partial t} = \sum_{i=1}^{2}\bar{\varepsilon}_i\bar{\delta}(x - x_i)\bar{R}_i(t). \tag{6}$$

Several universities and railway administrations have put together sophisticated computer programmes for advanced models in the present time.

It has appeared that the effects of moving axle loads may be expressed as the stochastic functions

$$F(t) = F + \varepsilon\dot{F}(t), \tag{7}$$

$$f(x,t) = [p + \varepsilon\dot{p}(s)][1 + \dot{r}(t)] \tag{8}$$

as derived in [8] and [9]. The schemes in Figures 13 and 14 represent a short span or long span bridge, respectively, corresponding to the equations (7) and (8). The calculations resulted in the estimations of the mean values and corresponding standard deviations for various cases.

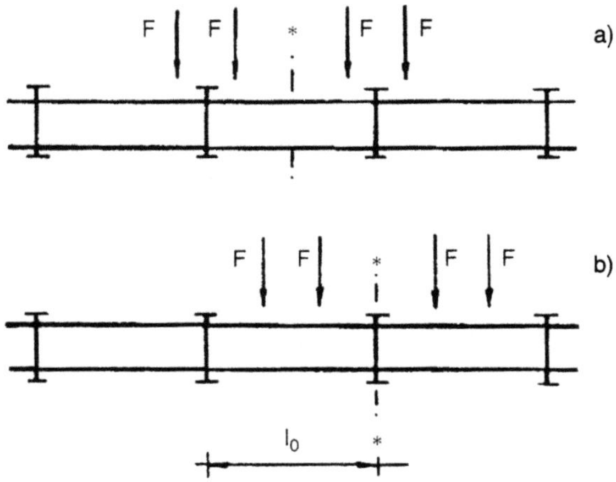

Figure 7. Cross beam effect due to axle forces F.

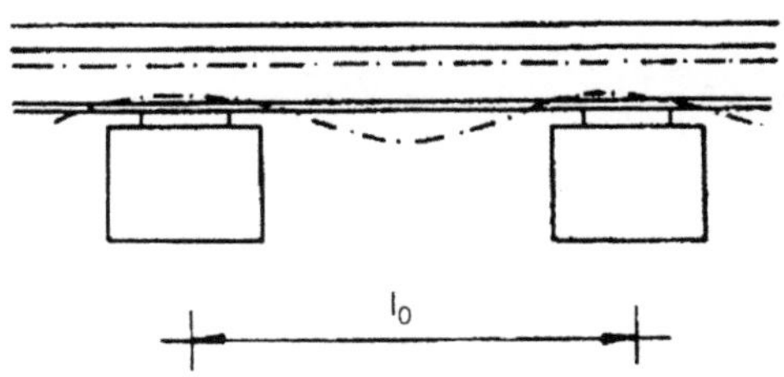

Figure 8. Sleeper effect.

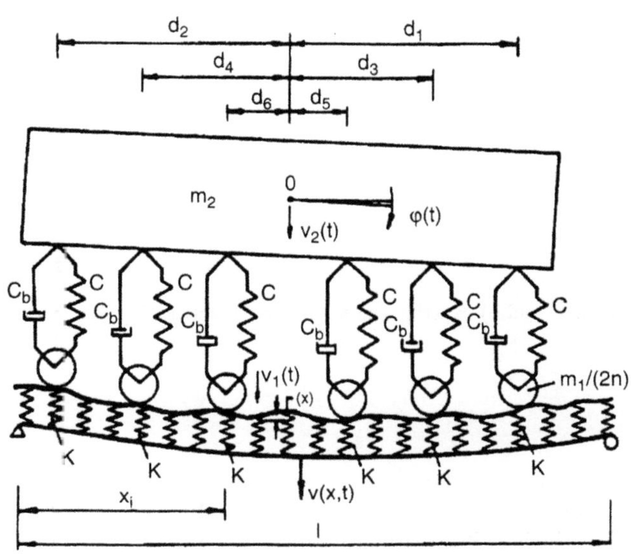

Figure 9. One of the first theoretical model of a locomotive with cross beam and sleeper effects.

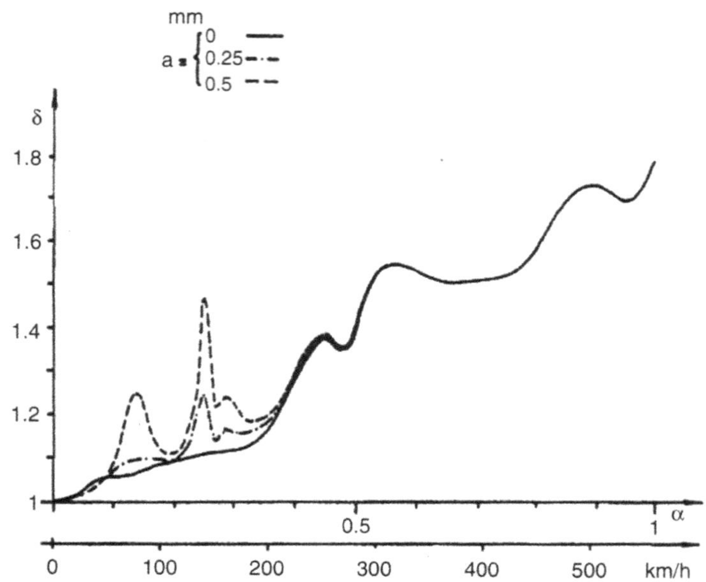

Figure 10. Dynamic impact factor δ as a function of speed c, sleeper effect at various unevenness, DB concrete bridge, $l = 10$ m.

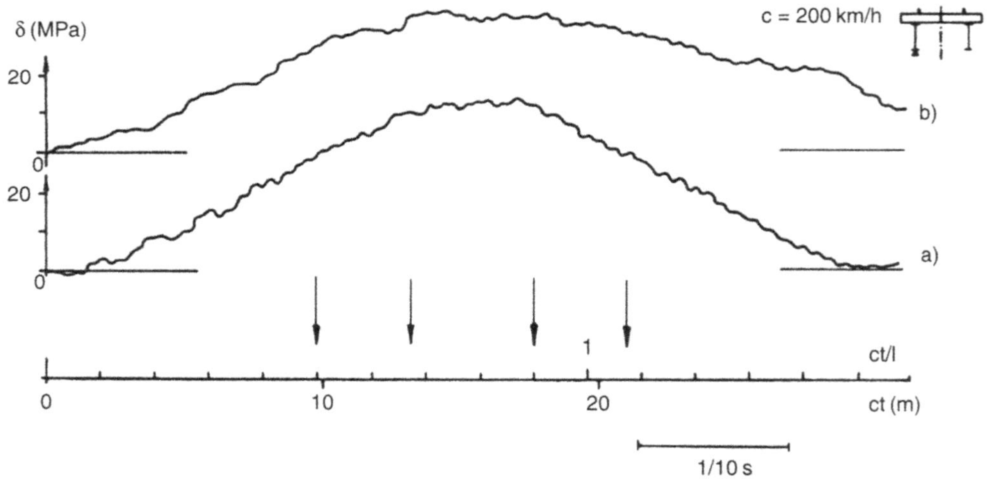

Figure 11. a) Theoretical and b) experimental stress-time histories, DB steel bridge, span 19.6 m, speed 200 km/h.

3 PRESENT

The speeds on railway lines have risen approximately by 100 km/h each 10 years in recent time so that they attain 350 km/h now in commercial passenger traffic but they reached the speeds higher than 500 km/h in experimental tests. However, the speed growth brought the increase of dynamic effects on bridges. The Figure 15, [10], shows an example of resonant vibration of a SNCF bridge under two TGV trains at speed of 192 km/h.

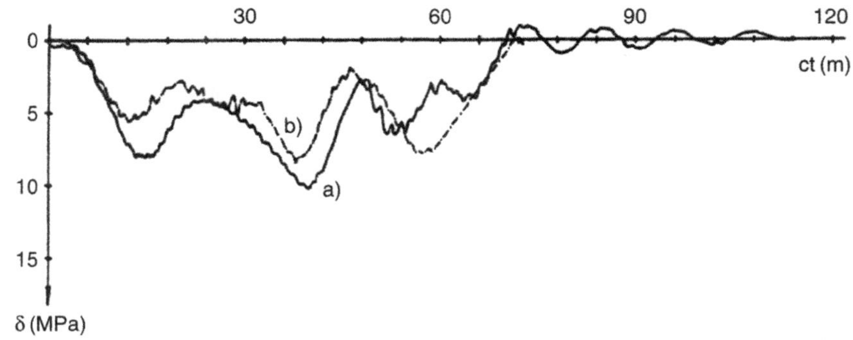

Figure 12. a) Theoretical and b) experimental stress-time histories, SNCF composite bridge, span 26.4 m, speed 241 km/h.

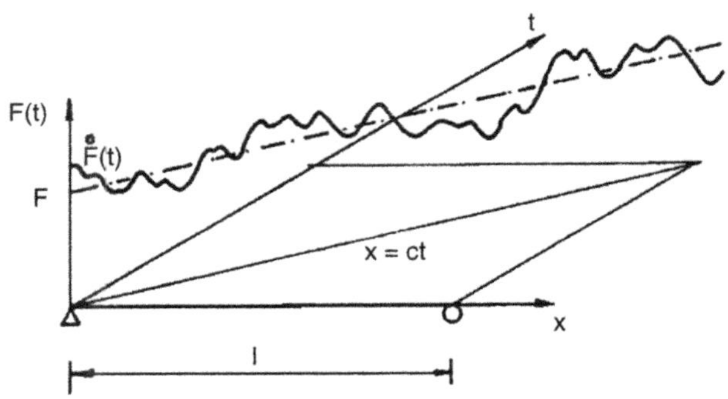

Figure 13. Movement of a random force along a beam.

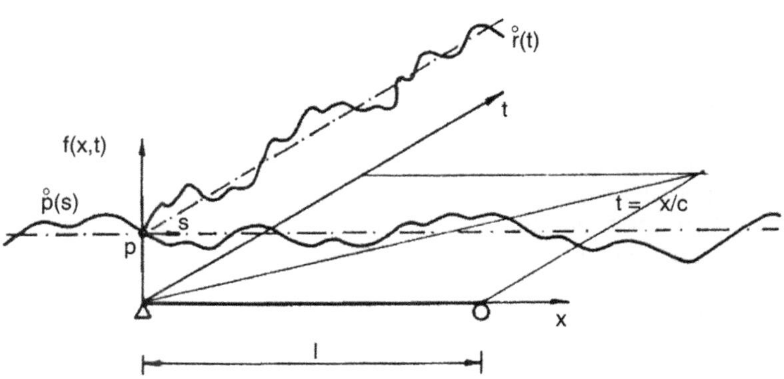

Figure 14. Movement of a random strip load along a beam.

The phenomenon may be idealized by the dynamic action of a row of axle forces and described by the equation

$$EI\frac{\partial^4 v(x,t)}{\partial x^4} + \mu\frac{\partial^2 v(x,t)}{\partial t^2} + 2\mu\omega_d\frac{\partial v(x,t)}{\partial t} = \sum_{n=1}^{N} \varepsilon_n(t)\delta(x - x_n)F_n \tag{9}$$

The disagreeable consequences of the intensive (resonant) vibrations are: the destabilization of ballast, risk of derailment, deterioration of passenger comfort and frequent maintenance.

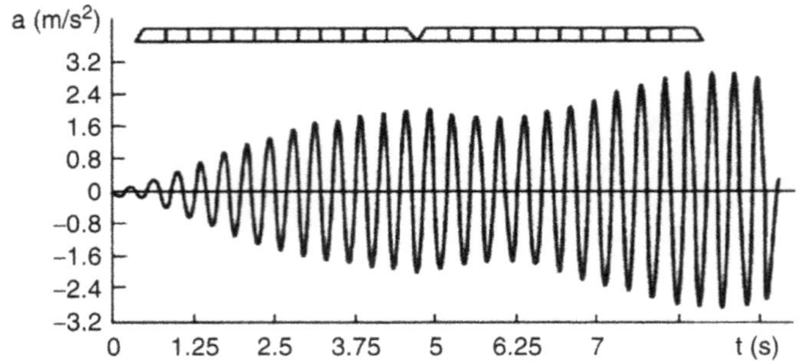

Figure 15. Resonant vibration of a SNCF bridge, span 38 m, under two TGV trains [10].

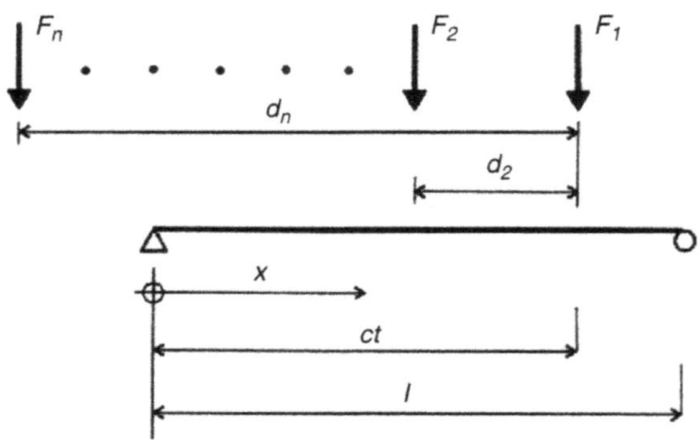

Figure 16. Movement of a row of axle forces along a beam.

Therefore, a huge effort was given to this problem at various ERRI questions. A simple model in the Figure 16 and equation (9) presented the deflection-, stress- and vertical acceleration-time histories together with stress range histograms at low and resonant speeds as shown in the Figures 17 and 18, [11], [15].The effects of the speed on deflections and bending moments are compared in the Figure 19, [15].

The vertical acceleration of bridges has appeared as a limit state at some cases as it affects the ballast loosening, [12]. The Figure 20 shows the effect of the speed on vertical acceleration of the bridge center. The allowed limit accelerations are 3.5 m/s^2 on bridges with ballast and 5 m/s^2 on ballastless bridges, respectively.

For fatigue, it is important to investigate the stress spectra which give a survey of appearance of various stress ranges, [13], [14], [15]:

$$\Delta\sigma = \sigma_{\max} - \sigma_{\min} \tag{10}$$

The counting method "rain-flow" was used for this purpose, [7], [15]. The Figures 21 and 22 show the schematic drawing of the counting procedure.

The rain-flow method results in histograms, the examples of which are depicted in the Figures 23 and 24, [13], [14], [15].

The effect of the speed on stress spectra is shown in the Figure 25, [15].

Various critical speeds appear at intensive vibration, [13]. The speed

$$c_{cr} = \frac{df_j}{k}, \quad j = 1, 2, 3, \ldots, \quad k = 1, 2, 3, \ldots, \frac{1}{2}, \frac{1}{3}, \frac{1}{4}, \ldots \tag{11}$$

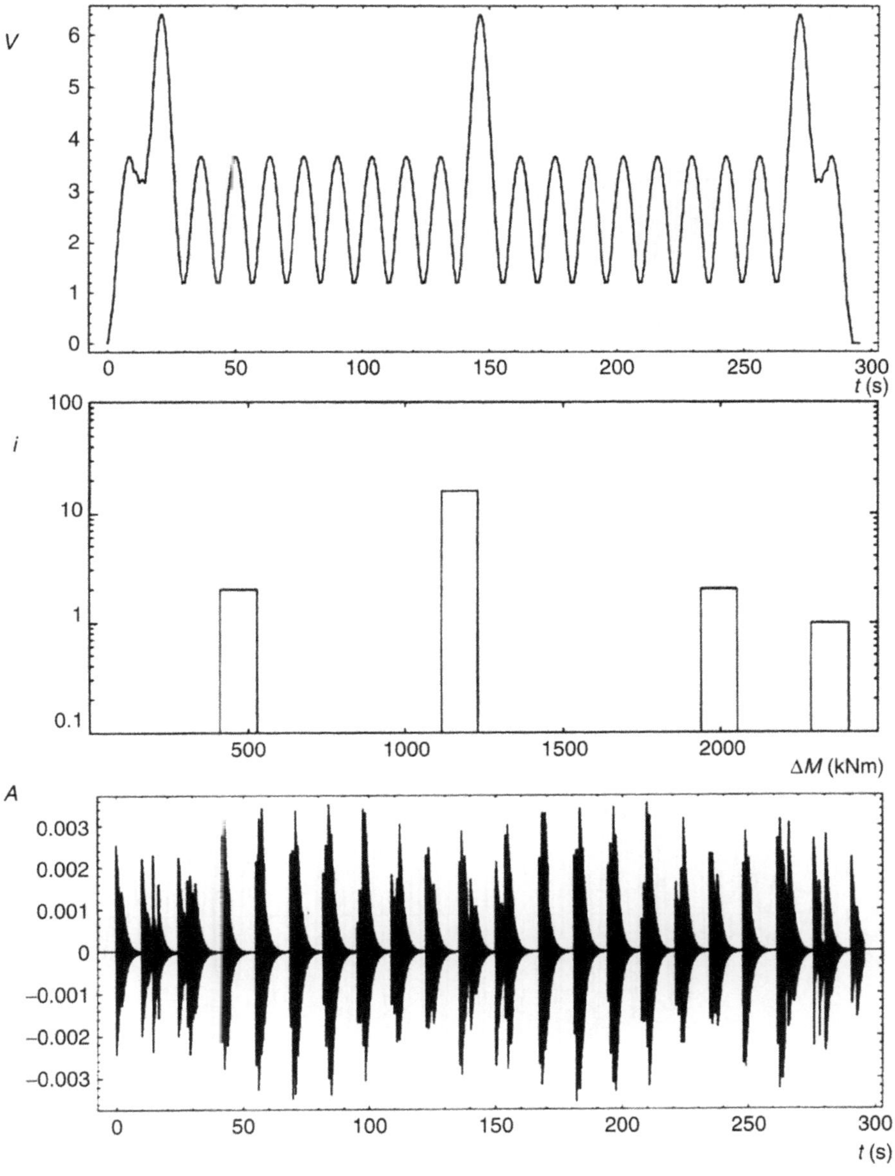

Figure 17. Time histories at low speed 5 km/h, steel bridge, $l = 20$ m, Eurostar.

is coming from the condition of repeated action of axles running in the equal distances d, while the speed

$$c_{cr} = \frac{2lf_j}{j}, \quad j = 1, 2, 3, \ldots \tag{12}$$

causes the destabilization of the bridge beam (an analogous phenomenon as stability). The critical speeds (11) appeared on some bridges of a few railways, however, the speeds (12) have not yet been reached in practice.

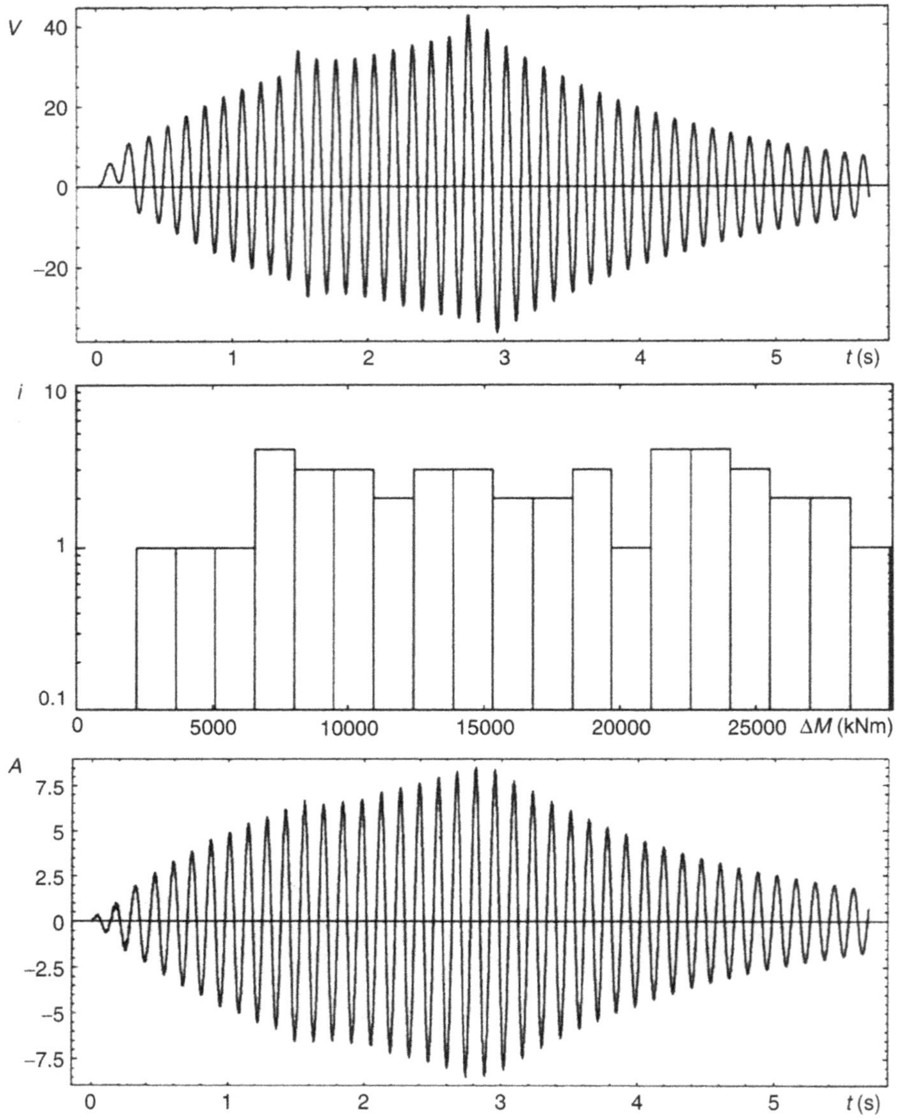

Figure 18. Time histories at resonant speed 500 km/h, steel bridge, $l = 20$ m, Eurostar.

4 FUTURE

The civil engineers have tried to diminish the dynamic response of bridges. For this purpose, a lot of systems have been invented. E.g., the elastic supports protect the bridges in seismic regions, [16], Figure 26, while the control system was suggested in [17] for the triangular falseworks, which is suitable for short spans, Figure 27.

In recent time, the double systems have been introduced where an elastic layer appears between two support systems. They are shown in Figures 28 and 29. The first system (double beam, [18]) is suitable for short and medium spans but it is expensive for long spans. The double string system, Figure 29, is too soft for railway bridges, [19].

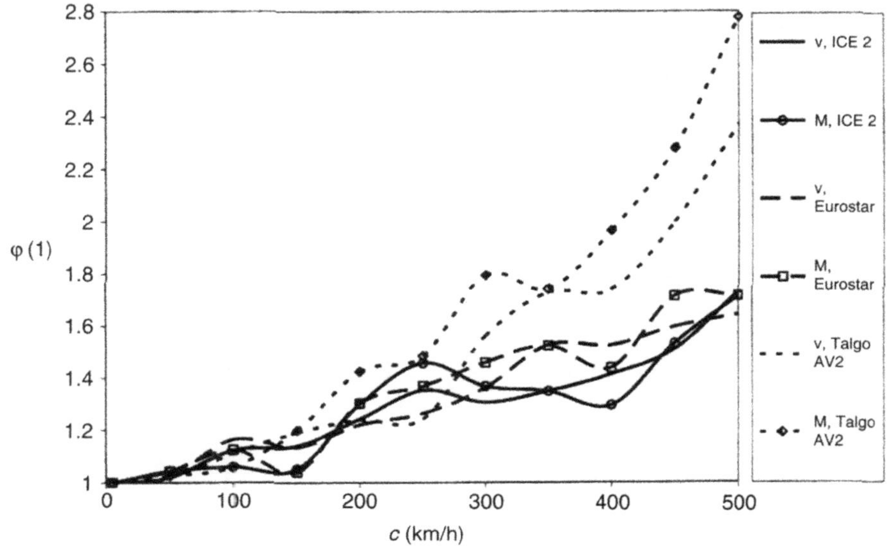

Figure 19. Effect of the speed on deflections and bending moments, steel bridge, $l = 5$ m, under various trains.

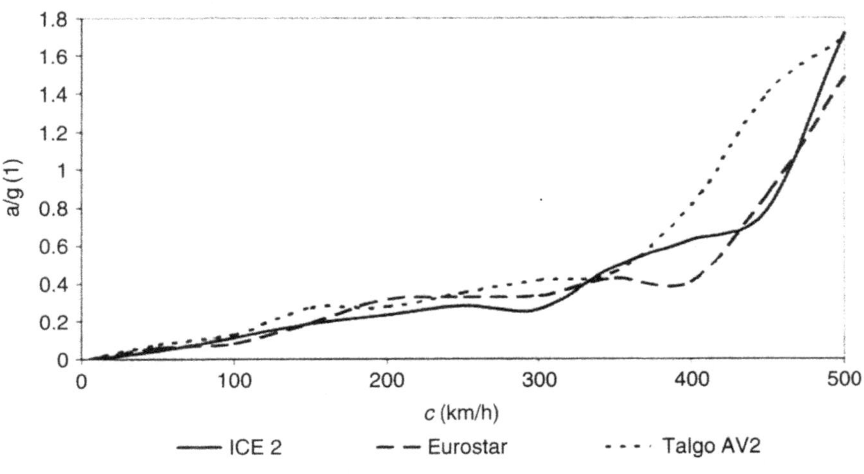

Figure 20. Effect of the speed on vertical accelerations, concrete bridge, $l = 10$ m, under various trains.

Therefore, a beam coupled with a string was suggested in [20], [21], which idealizes the pre-stressed bridges – the most spread system on railway bridges of short and medium spans. The Figure 30 represents a simple theoretical model and the equations

$$EI\frac{\partial^4 v_1(x,t)}{\partial x^4} - N_1\frac{\partial^2 v_1(x,t)}{\partial x^2} + \mu_1\frac{\partial^2 v_1(x,t)}{\partial t^2} + k[v_1(x,t) - v_2(x,t)] = \varepsilon(t)\delta(x - ct)F, \tag{13}$$

$$-N_2\frac{\partial^2 v_2(x,t)}{\partial x^2} + \mu_2\frac{\partial^2 v_2(x,t)}{\partial t^2} + k[v_2(x,t) - v_1(x,t)] = 0 \tag{14}$$

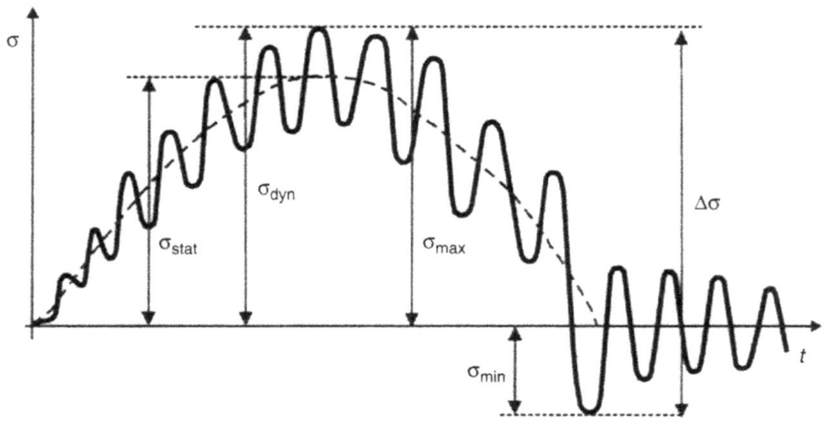

Figure 21. Stress-timer history, counting of stress ranges.

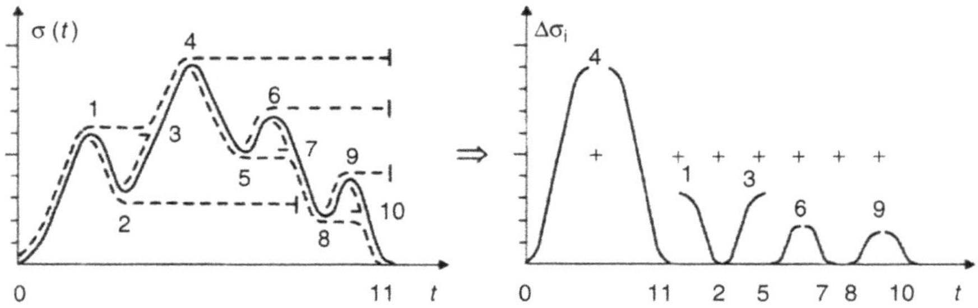

Figure 22. Counting of stress ranges, rain-flow.

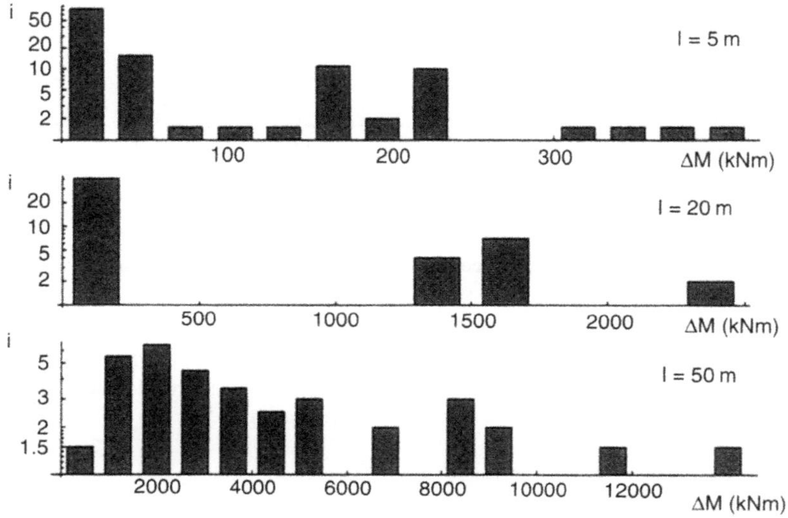

Figure 23. Stress spectra of steel bridges of spans 5, 20 and 50 m under ICE train, speed 350 km/h.

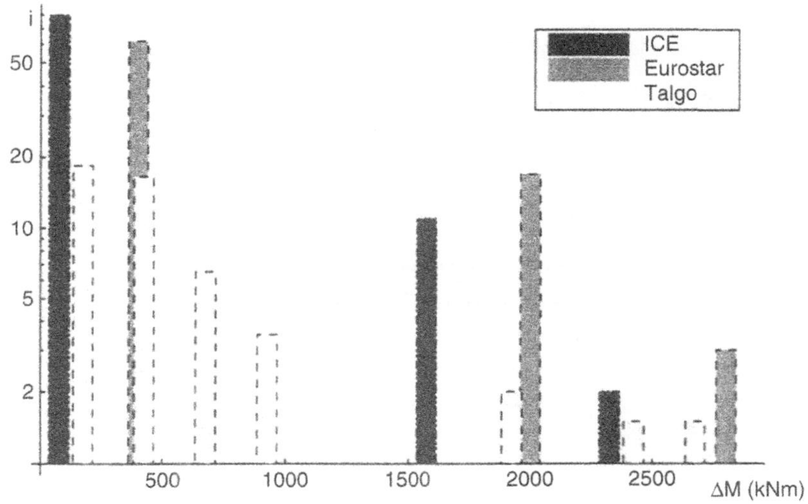

Figure 24. Stress spectra of a concrete bridge, span 20 m, under various trains, speed 350 km/h.

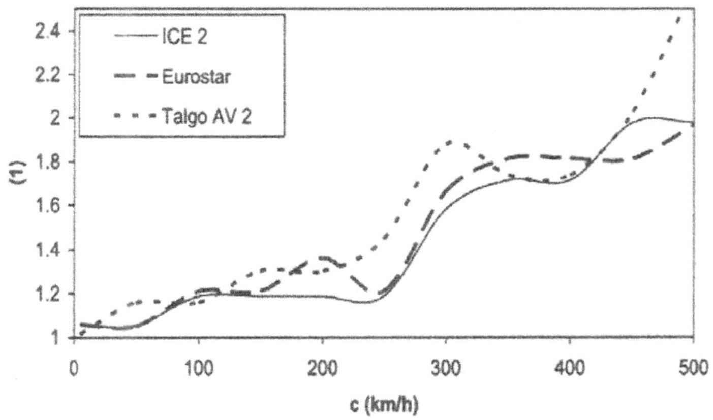

Figure 25. Effect of the speed on maximum stress ranges, concrete bridge, span 5 m under various trains.

Figure 26. Elastic supports of a bridge, [16].

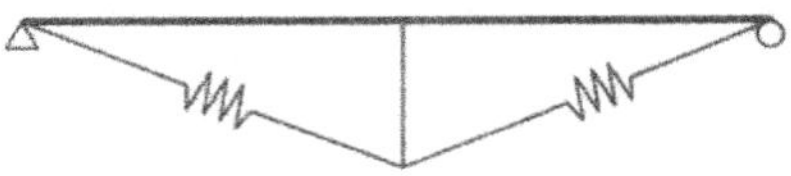

Figure 27. Triangular falsework system, [17].

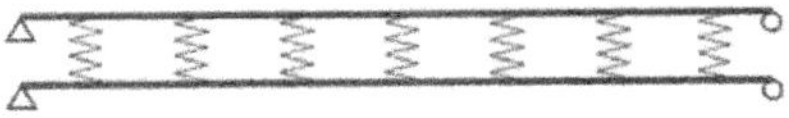

Figure 28. Double beam system, [18].

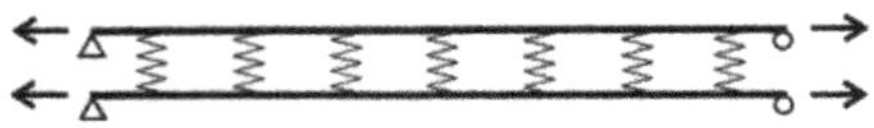

Figure 29. Double string system, [19].

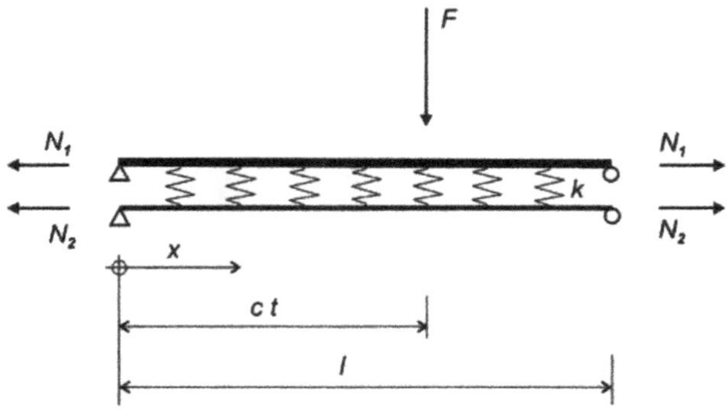

Figure 30. Beam coupled with a string subjected to a moving force F.

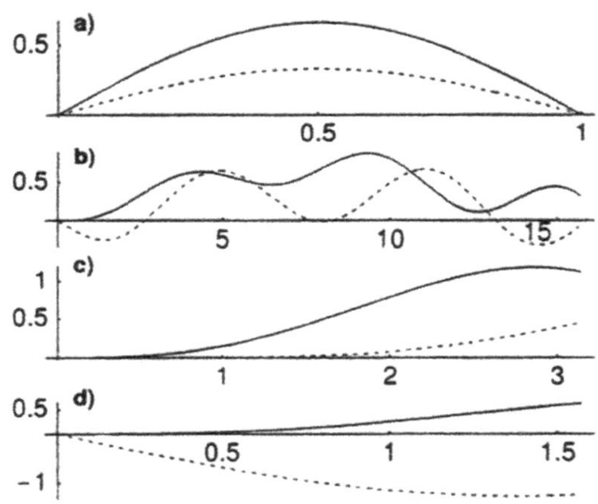

Figure 31. Deflection-time history of the beam and string at various dimensionless speeds a) $\alpha = 0$, b) $\alpha = 0.2$, c) $\alpha = 1$, d) $\alpha = 2$.

describe the dynamic behaviour of the beam and string, respectively, subjected to a single moving force.

The Figure 31 shows the deflection-time history of the beam (full line) and that one of the string (dashed line) at the centre of the system under the action of a moving force at various speeds, [20], [21]. The Figure 32 represents the effect of the speed, while the Figure 33 shows the space of three dimensionless parameters, where the effect of the slowly moving force on the beam deflection is expected lower than the quasistatic one.

In the distant future, the MAGLEV system may be introduced where the speeds will be higher than 500 km/h. The system will surely bring new problems, nevertheless, the basic principles will remain analogous to the equations (3) to (6).

5 CONCLUSIONS

Dynamic effects on bridges rise with increasing speeds of trains. The pertinent function provides several local peaks depending on the complex system bridge and train.

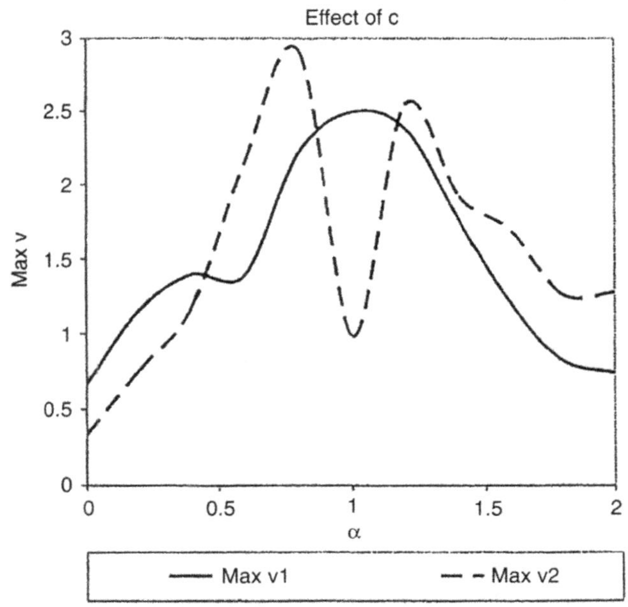

Figure 32. Effect of the speed on the coupled system beam and string, [20].

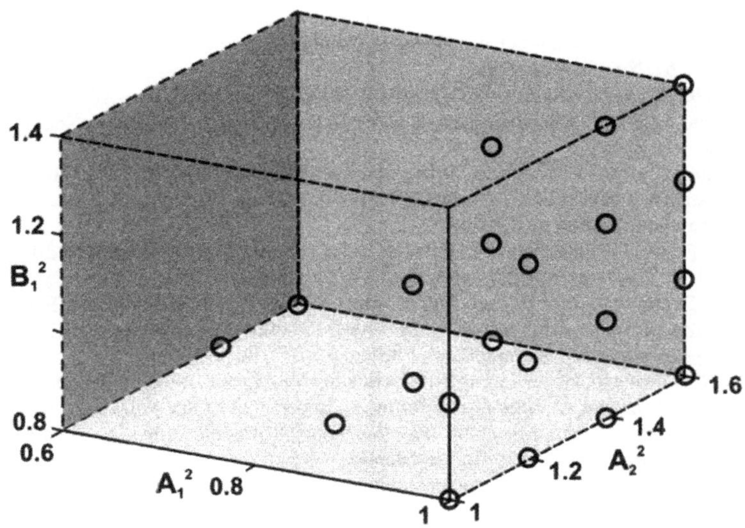

Figure 33. Space of dimensionless parameters A1, A2 and B1 where the low effects of the moving force on the beam may be expected, [21].

The dynamic response of concrete bridges is a little lower than steel ones due to their different mass and damping ratios.

Stress spectra present important data for the assessment of bridges at fatigue.

Vertical accelerations of bridges may be a limit state for the design of new high speed lines.

The diminishing of bridge response could be achieved with dampers or by interconnection of prestressed beams with pretensiled strings.

ACKNOWLEDGEMENT

The supports of the grants GA AS CR A 200710505, GA CR 103/05/2066, MOSTDYN 1F45D/013/120 and the research plan of ITAM AV OZ 07207/10524 are gratefully acknowledged. Thanks also to Mrs. RNDr. I. Frolíková for the arrangement of the paper presentation.

REFERENCES

[1] Willis, R. 1849. Preliminary Essay to Appendix B: Experiments for Determining the Effects Produced by Causing Weights to travel over Bars with Different Velocities". In: Grey at all: *Report on the Commissioners Appointed to Inquire into the Application of Iron to Railway Structure*. W. Clows and Sons, London.

[2] Stokes, G.G. 1849. Discussion of a Differential Equation Relating to the Breaking on Railway Bridges. *Transactions of the Cambridge Philosophical Society*. 8, part 5, p. 707–735.

[3] Timoshenko, S.P. 1922. On the Forced Vibration of Bridges. *Philosophical Magazine*, No. 6, Vol. 43, p. 1018.

[4] Inglis, C.E. 1934. *A Mathematical Treatise on Vibration in Railway Bridges*. The University Press, Cambridge.

[5] Koloušek, V. 1973. *Dynamics in Engineering Structures*. Butterworth, London.

[6] Frýba, L. 1999. *Vibration of Solids and Structures Under Moving Loads*. 3-rd. ed., Thomas Telford, London.

[7] Frýba, L. 1996. *Dynamics of Railway Bridges*. 2-nd ed., Thomas Telford, London.

[8] Frýba, L. 1976. Non-stationary Response of a Beam to a Moving Random Force. *Journal of Sound and Vibration*, Vol. 46, No. 3, p. 323–338.

[9] Frýba, L. 1977. Stationary Response of a Beam to a Moving Continuous Random Load. *Acta Technica ČSAV*, Vol. 22, No. 4, p. 444–479.

[10] Ramondenc, Ph. 1998. Vom Einfluss hoher Geschwindigkeiten auf den Entwurf von Eisenbahnbrücken am Beispiel der Stahlbrücken auf der Hochgeschwindigkeitsstrecke (HGS) des TGV Méditerrannée. *Stahlbau*, Vol. 67, No. 8, p. 652–658.

[11] Frýba, L. 2002. Intensive Vibration of Bridges due to High Speed Trains". In: J. Allan, R.J. Hill, C.A. Brebbia, G. Sciutto, S. Sone, J. Sakelari (eds): *Computers in Railways VIII*. WIT Press, Southampton, Boston, p. 595–604.

[12] Frýba, L. 2003. Vertical Accelerations of Bridges – a New Limit State. In: C.A. Brebbia, G.M. Carlomagno, P. Anagnostopoulos (eds): *Computational Methods and Experimental Measurements XI*. WIT Press, Southampton, Boston, p. 299–308.

[13] Frýba, L. & Fischer, C. 2003. Number of Stress Cycles in Bridges due to High Speed Trains. In: IABSE Reports, Vol. 87: *Structures for High-Speed Railway Transportation*. IABSE, Zürich, p. 68–69, CD Rom.

[14] Frýba, L., Fischer, C. & J.-D. Yau 2003. Stress Ranges in Bridges Under High Speed Trains" In: B.H.V. Topping (ed.): *Proceedings of the 9-th International Conference on Civil and Structural Engineering Computing*. Civil-Comp Press, Stirling, p. 185–186, CD Rom.

[15] Frýba, L. 2004. Dynamic Behaviour of Bridges due to High Speed Trains. In: R. Delgado, R. Calçada, A.C. Matos (eds): *Bridges for High-Speed Railways*. Proceedings of the Workshop, Porto, p. 137–158.

[16] Yang, Y.B., Yau, J.D. & Wu, Y.S. 2004. *Vehicle-Bridge Interaction Dynamics – With Applications to High-Speed Railways*. World Scientific, New Jersey.

[17] Reckmann, H. 2002. Aktive Schwingungsreduktion an einem elastischen Fahrweg unter bewegter Masse. *Fortschritte-Berichte VDI*, Reihe 11, No. 309, VDI Verlag.

[18] Kawazoe, K., Kono, I., Aida, T., Aso, T. & Eibisuda, K. 1998. Beam-type Dynamic Vibration Absorber Compromised of free-free Beam. *Journal of Engineering Mechanics*, p. 476–479.

[19] Oniszczuk, Z. 2000. Transverse Vibrations of Elastically Connected Double-String System", Parts I, II. *Journal of Sound and Vibration*, Vol. 232, No. 2, p. 355–386

[20] Frýba, L. & Fischer, C. 2005. Dynamics of Prestressed Beams Coupled with a String. In: C.A. Brebbia, G.M. Carlomagno (eds): *Computational Methods and Experimental Measurements XII*. WIT Press, Southampton, Boston, p. 445–454.

[21] Frýba, L. & Fischer, C. 2005. Vibration of a Coupled System Beam and String Under a Moving Force. In: C. Soize, G.I. Schueller (eds): *Proceedings EURODYN 2005*, Paris, in press.

CHAPTER 4

Experimental verifications of vibration – and fatigue conduct of viaducts and large span bridges, crossed by high-speed trains

P. Van Bogaert
Department of Civil Engineering, Ghent University, Belgium

ABSTRACT: Before commissioning and after some time of commercial service on the Belgian section of the high-speed line from Paris to Brussels, the actual train car accelerations were measured by a special test train. Close analysis of the measured values enabled to compare to predicted values. The measurements show in general 30% lower values of the predicted maximum amplitude of instantaneous accelerations. In addition, the measurements allowed verifying experimentally the galloping movements of trains crossing longer viaducts. This galloping may show a resonance frequency, increasing fundamentally the cyclic angular rotation about a horizontal axis of train cars. Fourier-analysis of the peak values of accelerations allows the definition of a resonance number. The main influence factor for this phenomenon is the ratio of the train speed to the individual span of the viaduct. The results of measurements were also compared to international recommendations. This comparison suggests that the recommendations are insufficient for long span bridges and 300 km/h high-speed bridges. Several steel arch bridges have been tested on full-scale level for stress concentrations in main elements and in orthotropic plated decks. The tests show that continuous measurement of stresses is indispensable for obtaining all peak values. In addition, the tests have revealed that, due to the ballast, the stress variations from railway loading are considerably lower than for road traffic. This demonstrates that steel orthotropic plated decks are a solution, which is particularly well adapted to railway traffic. At present, the development of the High-Speed network is reaching its final stage in Belgium and already new line extensions are being prepared. Various designs for these extensions are shown and explained in this paper.

1 INTRODUCTION

Railway vehicles, crossing long span bridges or multi-span viaducts produce much higher displacements and accelerations of bridge superstructures as when these loads are acting statically. These effects increase rapidly with speed and have become well-known in the design of high-speed railway structures. The consequences of the increase of displacements and indeed those of accelerations are twofold. Force resultants, stresses and deformations of bridge superstructures are increasing rapidly with train speed and loading models may become inaccurate or underestimate the effects of real train vehicles. In addition, the bridge deck accelerations may exceed values at which track ballast looses its cohesion and behaves as a liquid. On the other hand, deformations and accelerations of the superstructures may on their turn be transmitted to the train cars. Interference of the train car movement with these time-dependent quantities causes increasing of the resulting train car accelerations and discomfort to passengers. As a consequence, passengers can feel nauseated. This effect has been examined widespread and is included in recommendations of Eurocode EN 1990 Annex 2 [1]. In addition the limiting values for bridge stiffness and conditions in general are more critical than other effects of high-speed. Prediction of passenger comfort conditions is carried out mostly by applying a time-dependent series of axle loads with a saw-tooth scheme, as was proposed

Figure 1. Viaduct at Antoing – view from inside the structure.

since 1993 [2]. Modal analysis, followed by transient response calculation is then being carried out. As a result a time-dependent spectrum of bridge deformations or accelerations is found. These data are used as input for a multi-body system of train cars, with suspension system of springs, masses and dampers. The main results are the train car accelerations at the locations of the train car bogies. These accelerations are a consistent criterion for evaluating passenger comfort.

The aim of the present research was to evaluate measured values of train car accelerations in different circumstances and for several bridges and viaducts. The results have been confronted with calculated values and with some recommendations of Eurocode EN 1990 Annex 2. Values of train car accelerations have been measured on the viaducts of Antoing, Arbre, Lembeek and Lot and on the steel tied arch bridge of Halle. In addition, numerical simulations have been carried out for 3 bridges on the high-speed section between the Belgian-Dutch border and Breda.

Before discussing measurement results and conclusions concerning the dynamic behaviour, the various bridges and viaducts, considered during the research are being presented. Each time the ratio of the individual span to the maximum deflection caused by the scheme of Load Model 71 is given. This parameter allows the assessment of the general stiffness of the structure and is being used during further discussion.

2 EXPERIMENTAL VERIFICATION OF VIBRATIONAL CONDUCT

2.1 *Structures considered for verifications*

2.1.1 *Antoing viaduct*

Looking in the direction from Paris to Brussels, the structure consists of 5 concrete trough-section spans of 50 m, a steel-concrete span with steel arch of 116.8 m and a final 50 m long concrete span. The longest span allows the crossing of the river Scheldt and is shown from the inside in Figure 1.

Expressing the stiffness as the ratio of span to the vertical sag due to placing 2 loading schemes of LM 71 on the superstructure, the normal spans have $L/y = 4200$ whereas the longer span has $L/y = 3850$.

2.1.2 *Arbre viaduct*

The Arbre viaduct (Fig. 2) is the longest structure on the Belgian high-speed network. Its concept is similar to the smaller spans of the Antoing viaduct. The structure allows crossing a valley of compressible soft soil and several roads as well as a small river.

Figure 2. Viaduct at Arbre – view from the air.

Figure 3. Lembeek Viaduct: shorter spans.

2.1.3 *Lembeek viaduct*

The Lembeek viaduct (Fig. 3) is located at the very beginning of the 300 km/h speed section, some 23 km to the South of Brussels. It consists of trough section concrete single track elements. From South to North there are 4 × 19 m spans, 6 × 31.5 m spans and again 14 × 19 m.

The larger spans consist of parallel girders. The stiffness ratio of all spans is identical and equals L/y = 1300. The viaduct is particular in this sense that it is the location where the rail, current and signaling changes to high-speed technology. In addition, this viaduct is heavily blotted by the huge noise barriers, as it passes right through a village.

This viaduct and the next Lot viaduct are examples of using precast girders and elements, which is widely spread in Belgium. Precast beams with bonded pretensioning, having length up to 45 m are produced easily by concrete manufacturers.

2.1.4 *Lot viaduct*

The completely prefabricated steel-concrete viaduct of Lot (Fig. 4) [3] allows the crossing of the high-speed line Paris-Brussels with domestic tracks. Located 8 km to the South of Brussels,

Figure 4. Lot Viaduct consisting of precast HP-concrete and steel trusses.

Figure 5. Double steel tied-arch bridges at Halle during test loading.

it consists of 16 identical spans of 41.5 m. The piers are alternatively straight and triangular elements, whereas each bridge deck has two lateral beams of HP-concrete C 80/95 with steel encased beams. Steel truss structures are bolted to the steel encased beams. Their principal function is to increase the bending stiffness. The stiffness ratio L/y equals 1300.

2.1.5 *Steel tied arch bridges at Halle*

The steel tied arch bridges at Halle for the high-speed line and for domestic tracks are located 20 km to the South of Brussels. The structures consist of twin 115 m single-span arch bridges with orthotropic plated steel deck, both being linked by triangular hangers. The structures are similar to the bridge at Leuven [4].

The stiffness factor for these twin-bridges, one for domestic traffic, the second for high-speed trains, equals L/y = 1520.

2.2 *Measurements and results*

2.2.1 *Sampling of acceleration data*

The overview of section 2.1 mentions structures located on the Western Branch of the high-speed network in Belgium. The Eastern and Northern branches of the network have not bene commissioned yet. Hence, no experimental data are available at this time.

Figure 6. Recording of measurements by graphical output.

However, the East branch, to be opened in 2008, comprises 4 interesting concrete viaducts spanning the crossing of entrances to the adjacent motorway with the new high-speed line. In addition, this branch comprises the 'Hammerbrücke' viaduct across the river Gueule valley as well as the 'Prester'-bridge crossing a valley near the village of Lontzen. The Northern branch towards the Netherlands is to be commissioned in 2007 and comprises several bridges near Antwerp and two steel arch bridges, which are being discussed in section 5.2.

Because of this, most of the measured train car accelerations were recorded during test periods of the Lille to Brussels high-speed section. More recent measurements are heavily influenced by track irregularities and do not show a clear image of the influence of bridges and viaducts, since the latter are really small. For the Lot viaduct 20 test samples could be used, whereas for the Halle bridge, the Lembeek, Arbre and Antoing viaducts 8 series of test samples were available.

Train car accelerations were measured by a special train, generally used for track quality assessment. Apart from vertical car accelerations, the equipment of this train also records horizontal accelerations, track alignment and other parameters. The charts being produced are purely graphical and must be digitalised, by the process as described in [5]. Figure 6 shows the recording of accelerations by means of charts. More recent equipment allows digital sampling of data, and will be used in the near future.

In addition, Figure 7 shows 3 typical graphical outputs of vertical accelerations measured by the special train, while crossing the Arbre, Lembeek and Lot viaducts at train speeds of 300 and 210 km/h; Obviously, these raw data must be treated in order to eliminate the effect of pen thickness and inaccuracies. All measured train car acceleration curves had to be re-assembled to eliminate vertical accelerations due to the vertical track alignment and to locate the structures exactly on the graphical data.

In addition, the zero baselines must be determined exactly, requiring the use of statistical methods or double integration of the output signal. The relevancy of both these statistical methods has been verified by assessment of the maximum error. Comparing to the maximum accelerations, the inaccuracy of the method is limited to 1.8%.

Additional calculations are concerned with deriving for each test graph the series of maximum and minimum train car accelerations. Fourier transform of the graphical output enabled to assess clearly the main frequencies and accelerations. Finally, the most characteristic values of vertical accelerations are determined as 95% fractal values of the total sample population.

2.2.2 *Steel tied arch bridges at Halle*

Unfortunately most trains cross the bridges being considered at their usual speed. Therefore, it is not possible to make sufficiently correct assessments of the variation of maximum and minimum

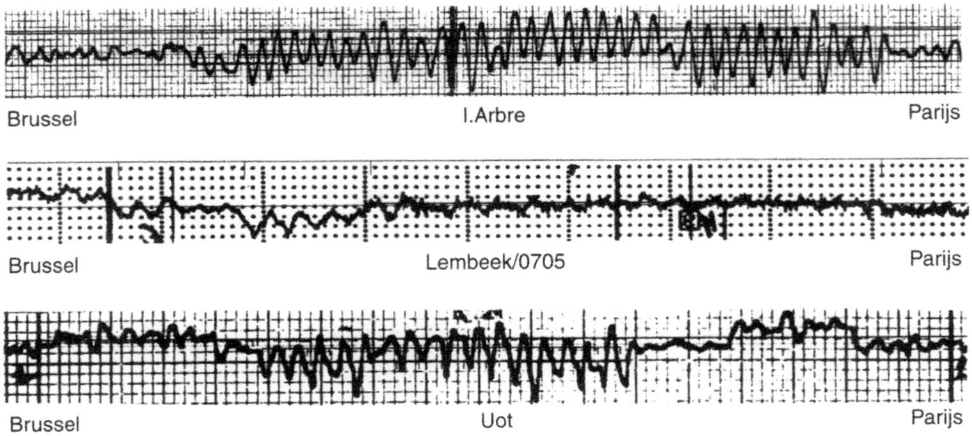

Figure 7. Typical graphical outputs of vertical accelerations on Arbre, Lembeek and Lot viaducts.

Table 1. Maximum acceleration values for Antoing and Arbre viaducts.

Viaduct	Antoing		Arbre	
	L = 50 m	L = 120 m	L = 50 m	L = 60 m
a_{max} (m/s^2)	0.789	0.489	0.566	0.713

Table 2. 95% Fractal acceleration values.

Viaduct	Halle HSL bridge		Lembeek viaduct	
Speed range	151–152 km/h	210–215 km/h	146–152 km/h	191–216 km/h
$a_{95\%}$ (m/s^2)	0.526	0.573	0.341	0.492

accelerations as a function of train speed. Both for the Antoing and Arbre viaducts, a relevant number of values is available to determine 95% fractal values of the acceleration extremes at v = 300 km/h. These are shown in Table 1.

For the viaducts of Lembeek (31.5 m spans) and the Halle bridge, sufficient values are available to distinguish 2 statistical speed classes, namely round 150 and 210 km/h. These results are shown in Table 2.

Since there are sufficient individual values, the 95% fractal value, below which only 5% of experimental results can occur, were taken. For the Lot viaduct, there are sufficient data to establish a fitting curve as a function of train speed. This has been derived as:

$$a = 4.4574 \cdot 10^{-4}\, v^2 - 0.022287\, v + 0.53777\, (\text{in m, s} \quad 100 < v < 220\, \text{km/h}) \qquad (1)$$

and corresponds to a second degree parabola, having horizontal tangent for v = 90 km/h.

2.2.3 *Galloping of train cars*

As the train cars are crossing the structure, the vertical accelerations of the front and rear suspension bogies are unequal. Consequently, the train car rotates about a horizontal axis. This is the well-known galloping movement of high-speed trains. While interpreting the measurement results, galloping was found regularly. However, the galloping pattern may be difficult to find. It is expected that

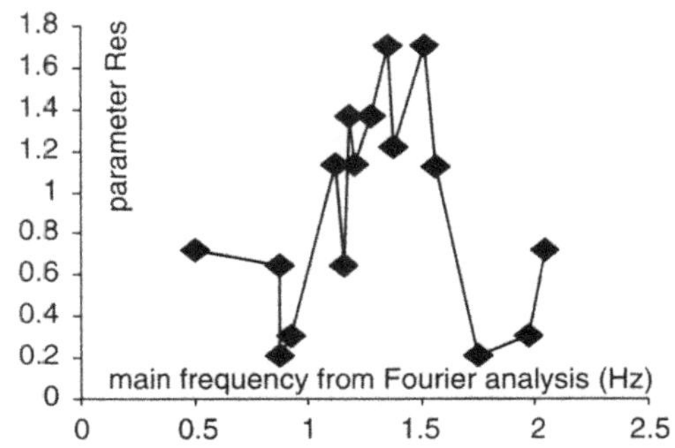

Figure 8. Resonance curve of galloping train cars.

the galloping frequency corresponds to the structure's span. The frequency then becomes $f = v/L$. To derive a relevant standard for the galloping effect, the accumulation of rotational energy has to be quantified. Looking at the measured accelerations, peak values may be easily identified. If 3 consecutive peak values are considered, and the third peak value C is larger than the values corresponding to peaks A and B, then the quantity p is defined as:

$$p_i = a_{\text{peak C}}/a_{\text{peak A}} \tag{2}$$

If the acceleration in C is lower than in A, pi is taken 0. This process is programmed for each series of measurements, and results in determining a resonance number as:

$$\text{Res} = \frac{\sum_{i=3}^{n} p_i^+ + \sum_{i=3}^{n} p_i^-}{2n-4} \tag{3}$$

Negative (downward) as well as positive (upward) peaks need to be considered in. The newly defined quantity Res is an accurate standard to evaluate the accumulation of kinetic energy. During Fourier-analysis, the instantaneous accelerations and velocities can be expressed as:

$$a(t) = \sum_i A_i \cos(2\pi f_i t) \quad v(t) = \sum_i \frac{A_i}{2\pi f_i} \sin(2\pi f_i t) \tag{4a,b}$$

Since kinetic energy is related to the square of velocity, by integration of the square values of eq 4(b) a standard for the energy of the several Fourier-terms can be found from:

$$E \approx A_i^2/f_i^2 \tag{5}$$

This quantity is related to Fourier transform. The data related to all structures have been assembled to the graph of Figure 8, showing the value of Equation (3) versus the main frequencies of the Fourier analyzed signal. As the main frequency is closer to the train car suspension frequency, the quantity from Equation (3) exceeds 1. This shows that galloping energy is being accumulated.

For any combination of v and L, the graph of Figure 8 allows determining whether train cars will actually show the galloping movement, causing discomfort to passengers. Clearly, the data for this graph relate to the suspension system of Thalys-train cars, used during all measurements.

2.2.4 *Comparing with recommended values*

According to EN 1990 Annex 2 paragraph A.2.4 4.3 limiting values of bridge deck stiffness may be expressed as L/y, y being the vertical deflection of the bridge deck, caused by LM 71 multiplied

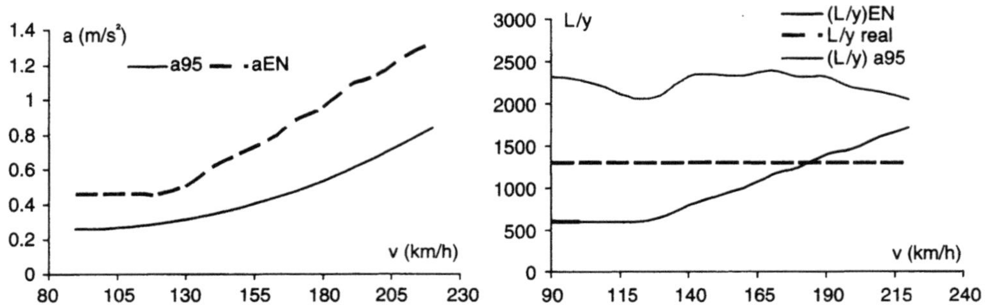

Figure 9. Comparison of 95% fractal accelerations and recommended values from EN 1990.

Table 3. Comparison accelerations for Halle bridge.

Train speed (km/h)	a_{95} measured	L/y_{EN}	L/y_{95}
150	0.526	600	1140
210	0.565	700	1238

with the appropriate dynamic factor Φ. The vertical acceleration then is inversely proportional to L/y. Two types of quantities can be determined to compare with the recommendations of EN 1990 Annex 2. The 95% fractal value a_{95} may be considered for a combination of L and train speed v. The value $(L/y)_{EN}$ corresponds to the recommended value to obtain maximum vertical acceleration of 1 m/s^2. Since accelerations are inversely proportional to L/y, a predicted value of a_{EN} may be found by the relation:

$$a_{EN} = (L/y)_{EN\ 1m/s^2}/(L/y) \tag{6}$$

Another comparison is made by determining the relative stiffness factor, corresponding to the 95%-fractal acceleration

$$(L/y)_{a95} = (L/y)_{EN\ 1m/s^2/a95} \tag{7}$$

For the Lot viaduct data for wide variety of train speed are available. Both quantities are compared in Figure 9. Obviously, the viaduct shows improved behaviour compared to the recommendations of EN 1990 Annex 2, the difference becoming clearer as speed is increasing.

The relative stiffness of this viaduct $L/y = 1300$ is certainly accurate. The value of the Equation (3) quantity found for this viaduct is below 0.5. To reach vertical accelerations of 1 m/s^2, the actual train speed would need to be 240 km/h. Should the values of EN 1990 Annex 2 have been adopted the required relative stiffness would be $L/y = 2025$, and the viaduct was to be more costly and more heavy.

The viaducts of Lembeek, Arbre and Antoing may be compared to the recommendations for singular train speed values only. In general, the recommendations of EN 1990 Annex 2 overestimate the required stiffness of all of these structures. The compared values for the Halle bridge are shown in Table 3, for two representative classes of train speeds. As the stiffness factor equals 1520, the table shows that EN 1990 Annex 2 requires the values of the last column to find the 95% value of the acceleration. Since the values from this column are much lower than 1520. The conclusion must be that EN 1990 Annex 2 underestimates the required stiffness for larger bridges. This is probably because larger bridges were little considered in the recommended values. A more complete comparison for the Halle bridge is reported in [5].

With reference to the curves of acceptable deformations of EN 1990 Annex 2, the graph of Figure 10 displays the chart taken from this recommendation and also showing the discrete

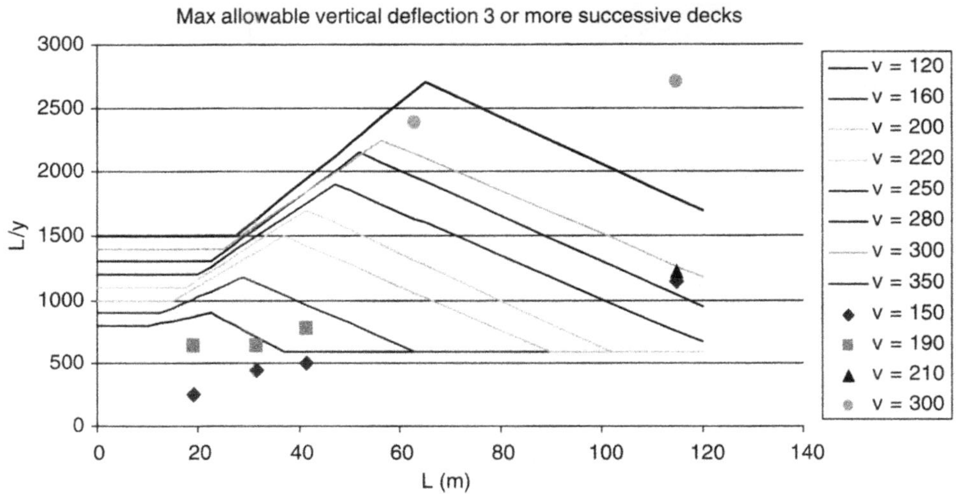

Figure 10. Comparison with recommended values from EN 1990 Annex 2.

combinations found for the viaducts being considered. This again confirms the finding aforementioned.

3 PREDICTIONS CONCERNING SHORT MASSIVE BRIDGES

3.1 *Short bridges and vibration conduct*

Several shorter bridges have been built close to the Belgian border on the high-speed section in the Netherlands. Most of these bridges consist of concrete slab decks, either multi-span continuous or single span. The viaducts mentioned in 2.1 all have a relatively large stiffness with respect to the cross section area, thus creating stiffer than massive structures. Short slab bridges are rather disadvantageous for vibration conduct, since the heavy mass provides low natural frequencies in spite of the short spans.

The examples shown hereafter are mainly concerned with requirements for the accelerations introduced in bridge superstructures, rather than passenger comfort. These requirements relate to track and ballast stability.

3.1.1 *Bridges being considered*

The first type of bridge is a 4-span continuous slab of 19.18 m – 2 × 23.5 m – 19.18 m spans. The slab has exactly 1 m thickness and is 14.38 m wide. Two additional corbels of variable depth from 1 m to 0.245 m and 1.39 m wide, carrying the walkways, are fixed to the slab. Two identical structures allow the crossing of the rivers Aa and Weerijs.

Each line of supports comprises 8 bearings of neoprene, consisting of 4 groups of 2 bearings. The distance between each group equals 3.9 m whereas the two bearings of one group are 0.7 m apart. The skew angle reaches 62.422°. This skewness should have a positive influence on the vibration conduct of the superstructure. The bridges are supported by intermediate solid piers and small abutments on a series of piles. The piers also have pile foundations. These bridges were cast in situ and demonstrate the massive nature of slab decks.

The bridge of Rhitse Street crosses this local road by a single span of 20 m and constitutes an almost straight crossing. The slab thickness equals 1.3 m whereas the width is 19.5 m.

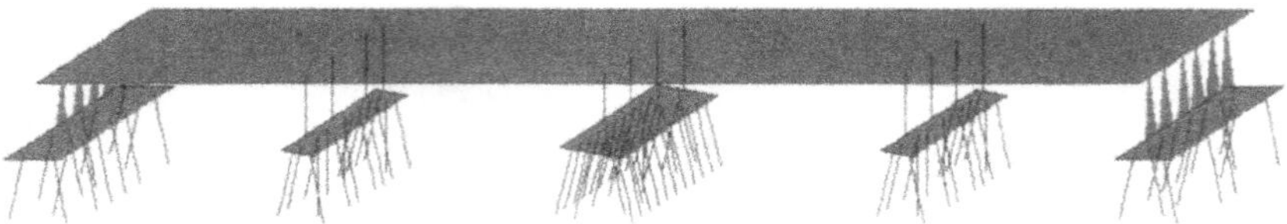

Figure 11. Complete model for dynamic behaviour of Aa and Werijs bridges.

Figure 12. Slab model for dynamic behaviour of Aa and Werijs bridges with ballasted track.

The superstructure is supported by 6 neoprene bearings at a distance of 2.25 m and placed on heavy abutments and direct foundation.

3.1.2 *Numerical methods*

The first model used for the Aa bridge, consisted of a complete structure, including the bearings and foundation piles. It is shown in Figure 11, including the types of elements used. Modal analysis of this type of model creates the problem of the lateral displacements of the piles. Vibration modes due to the unrestrained piles are however inapplicable in the real structure. This will be demonstrated in section 3.1.3. Hence, a less complicated model has been used for both bridges across the Aa and Rithse Street, consisting of slabs only. In addition, if ballast is used on the slabs, the concrete mass is increased virtually, thus taking into account the fact that the ballast should follow the displacements of the concrete slab. The concrete deformation modulus is taken the constant value of 31000 MPa.

3.1.3 *Relevant natural frequencies*

The lowest relevant natural frequency of the model including the infrastructure reaches 4.3585 Hz for ballasted track and 4.6202 Hz for unballasted track. The mode shape for the slab model of ballasted track is shown in Figure 12.

In contrast with the complete model, the lowest natural frequencies equal 3.3479 Hz for ballasted track and 3.5151 Hz for unballasted track. The discrepancy between results from different models demonstrates the difficulty of predicting dynamic behaviour. The elaborated model is inaccurate for describing the superstructure alone, whereas the simplified model underestimates natural frequencies.

3.2 *Instantaneous vertical accelerations of bridge superstructures*

3.2.1 *Ballasted track*

Continuing the use of slab models, the modes up till 30 Hz frequency have been considered during the dynamic response of the crossing of double Thalys-train vehicles. A time step of 0.001 s has been used and the modal damping was set at 0.02. The train speed was varied from 300 to 361.2 km/h by steps of 3.6 km/h. At each moment in the series of time steps, the upward and downward accelerations of each mesh point of the bridge deck are kept in an appropriate file. Subsequently, all maximum values of both types of accelerations, regardless of actual time step are represented on

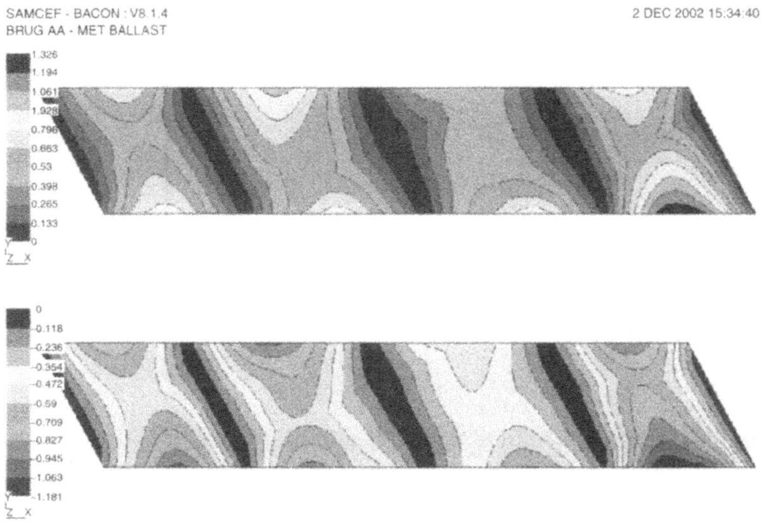

Figure 13. Peak values of upward (up) and downward maximum accelerations.

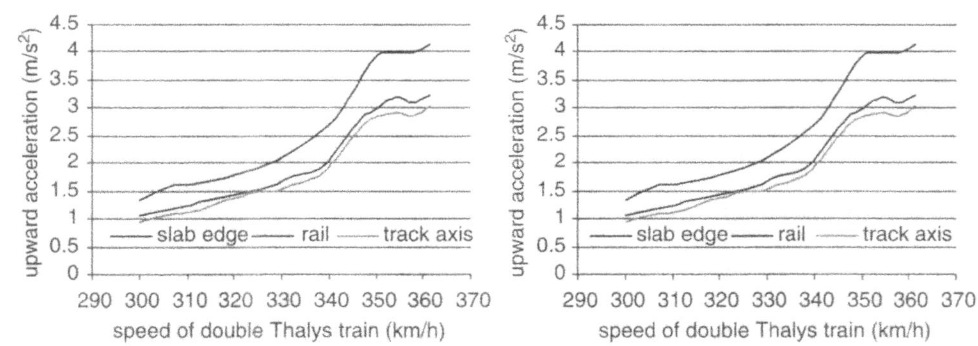

Figure 14. Peak values of slab accelerations as a function of train speed.

the slab deck as in Figure 13, for a certain train speed, in this case 300 km/h. This allows visualising an envelope of all peak accelerations during the crossing of a double Thalys train.

From Figure 13 appears clearly that the largest accelerations are found at the slab edges. These are irrelevant for ballast stability and also for the train conduct. This has been summarized further in the graphs of Figure 14. Obviously, up to 360 km/h train speed, the requirement of not exceeding an instantaneous acceleration of 3.5 m/s^2 is met for the track itself, albeit the slab edges may show higher accelerations.

The track designers however have used beam models to verify these bridges and found unacceptable accelerations, up to 4.3 m/s^2 at a speed of 308 km/h. They related this to the notion of critical speed:

$$v_{crit} = f_{e,n} \cdot 18.7 \tag{8}$$

In Expression (8) $f_{e,n}$ is the nth natural frequency, whereas 18.7 m is the distance between train bogies. According to our calculations the critical speeds should be those mentioned in Table 4.

3.2.2 *Ballastless track*

This analysis has also been carried out for the case of ballastless track. The accelerations appear to be some 15% higher. However, the requirements for track stability of unballasted track allow to

Table 4. Critical speed (km/h)

n	1	2	3	4	5
f_e	3.3479	3.5057	4.2594	5.3488	6.0658
v_{crit}	225.4	236.0	286.7	360.1	408.3

have accelerations of 5 m/s^2. Hence, the track designers decided to install this type of track on these bridges, as well as on the bridge type of Rithse Street, in spite of the fact that the lowest natural frequencies equal 4.76 Hz and 9.29 Hz, corresponding to critical train speeds of 320.7 km/h and 625.4 km/h.

4 EXPERIMENTAL VERIFICATION OF FATIGUE RESISTANCE

The establishment of the high-speed railway network in Belgium has required the building of 7 steel bridges of spans varying from 86 to 136.2 m. All of these are finished, although some important ones are still to be commissioned. As for all important structures, before commissioning the tracks thorough evaluation is necessary. This is mostly done by test loading. From this testing appears clearly that punctual positions of the test load do neither reveal the relevant maximum stresses, nor the stress variations, which is of particular interest with respect to the fatigue resistance.

4.1 *Global stress variations*

For the two high-speed steel arch bridges at Halle and Leuven a test-setup has been used, which enabled to verify the stresses and their variations in the lower chord, hangers and arches. The Leuven bridge, shown in Figure 15, has a span of 117 m, total width of 12 m, while the arch rise is 18.75 m. The hangers, connecting both the arches with the lower chord of the steel deck are arranged as triangles, thus increasing the bridge stiffness. The latter is highly recommendable, since in its final position, the bridge carries two railway tracks with a bend of 1200 m radius, the railway loading thus showing important eccentricity towards the bridge axis. More detailed information about the Leuven Bridge can be found in [4]. The steel plated bridge deck consists of an orthotropic structure, allowing minimum construction depth or maximum headway clearance for the subjacent canal. Clearly, the exact behaviour of the orthotropic plated deck, both for road and railway vehicles, has been a major occupation.

The test setup comprises 122 strain gauges of short types 3.05 to 6.35 mm, placed on the arches (see Fig. 16 left) and on the steel plated orthotropic deck (Fig. 16 right). During a first test phase, the bridge was loaded by heavy 44 tons lorries, constituting a total mass of 718 tons.

During this phase several load positions of the trucks were used, the most important being full load and 2 types of chessboard positions. As the actual structure was designed, using a full FE-model of the bridge deck and arches. Hence, calculated stresses may be compared to measured values. Obviously, the lorries were placed statically at closely defined locations. Taking the relevant stresses in the lower chord, arch springs and shortest as well as longest hangers, the comparison reveals in general that calculated stresses are larger than the measured ones. Figure 17 shows this comparison for full loading. The first series of stresses concerns the lower chord whereas the second are arch stresses (compression) and the third are hanger stresses. The average ratio of measured to calculated stresses equals 0.828 with standard deviation of 14%. However, stepwise loading of a structure does not guarantee to find the most relevant stresses. Bearing this in mind, during a second phase and tracks having been placed on the bridge, a new test loading by freight trains was organized. All strain gauges appeared to function well. In addition to this, two devices were installed for accurate location of train axles. An infrared device allowed, by interruption of the beam

Figure 15. Steel tied arch bridge at Leuven.

Figure 16. Strain gauge test setup on arch, lower chord and orthotropic plated deck.

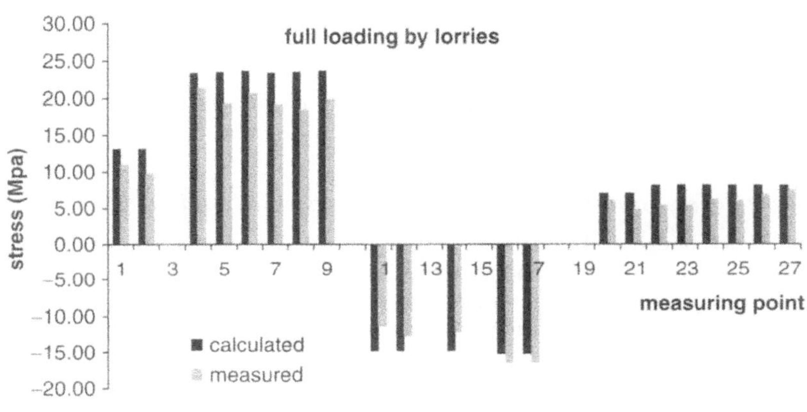

Figure 17. Comparison measured/calculated stresses during full loading Leuven bridge.

Figure 18. Train loading on Leuven bridge.

to identify the exact time at which axles are crossing. Simultaneously a laser beam was installed enabling accurate location of the axles.

The train load consisted of two 6-axle diesel locomotives of 106 tons each and 12 four-axle wagons, loaded with ballast and having masses varying from 69.2 tons to 77.1 tons. Figure 18 shows this train as it crosses the bridge on the curved track. At the time of the test, tracks were installed continuously at the Brussels' side of the bridge, but ended some 500 m at the Leuven side. Hence, the test train was unable to travel at speed larger than 7 km/h. For this loading, static positions of the train have been measured, as well as continuous stress measurement during the crossing. For the static measurements the average value of the ratio of measured to calculated values equals 0.937 with a standard deviation of 44%. Again the deviation of the results appears rather important, although basically it is due to some of the lower values.

If continuous measurements are analyzed, some conclusions become different. At times, the strain gauge signal becomes rather capricious including so-called disturbances. These may be due to insufficient filtering of the electric signal, which is rather improbable. More plausible reasons can be found in local effects and accelerations even at the low speed of the test train. This can be seen in Figure 19, showing the comparison of the continuously recorded stress at the lower flange of the arch with calculated values. Magnification of Figure 19 demonstrates that the peak values comprise 2 fractions. The basic fraction, appearing regularly and showing constant stress fluctuation, is obviously due to local loading effects and is missed in static measurement. The second fraction, appearing half as frequently as the first one, in an irregular manner, must be caused by local vibration effects. It appears clearly that static measurements overlook all of these and may seriously disregard relevant stress peaks.

4.2 *Behaviour of steel orthotropic plated bridge deck under railway loading*

Orthotropic plated bridge decks are heavily prone to fatigue damage since they contain various welded details where stress concentrations occur. Reducing the concentration effects in one detail generally increases the damage in most other locations. Hence, equilibrium for all effects has to be found. During the years the writer has been designing bridges with orthotropic plated decks, the improvements made have been such to develop a type particularly interesting for railway loading. This is demonstrated by the stress variation tests, carried out on several bridges.

The critical details for fatigue are the lower soffit of longitudinal stiffeners, the stiffener to deck plate welds and the cope hole edges of crossbeam cut-outs for continuous stiffeners through the

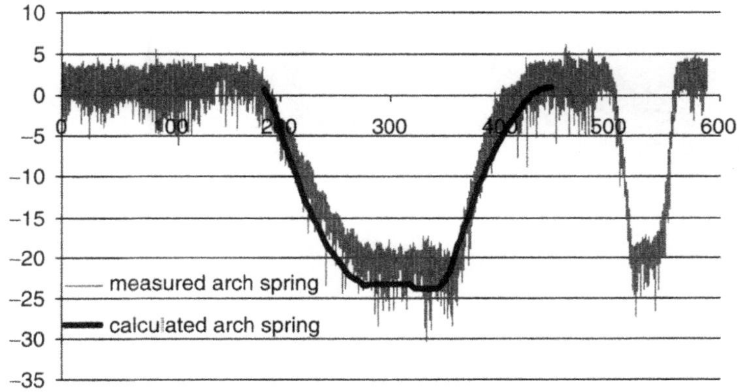

Figure 19. Continuous measurement of arch spring stresses.

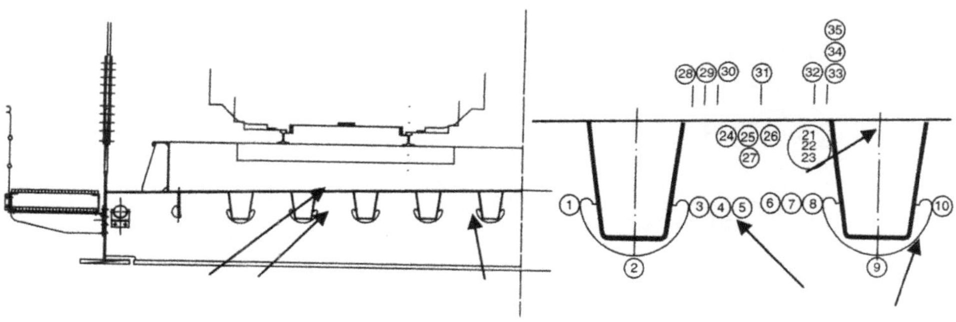

Figure 20. Half cross-section of orthotropic plated deck of railway bridge and detailed view.

crossbeam webs. These are shown by the arrows in the half cross-sections of the Leuven bridge in Figure 20 left and the detailed view of Figure 20 right.

In the appropriate code documents, these fatigue details are classified as classes 160 MPa, 80 MPa, the cope hole edge not being classified, according to present knowledge. However there should be no misunderstanding that these fatigue classes relate to nominal stress variations and not to hot spot stresses.

Measurements show that the orthotropic action under rail loading is very poor. This means that the concentrated wheel loads do not distribute well among the longitudinal stiffeners. This is due to the poor effective torsion stiffness as explained in [6]. Hence, the lower soffit stress variations are important for railway loading. This is why cope holes are used in the crossbeam webs, to improve the fatigue class of the stiffener soffit at this particular section. However the cut-outs and additional cope holes weaken heavily the crossbeam web plate, in particular if shear forces occur. This can be seen in Figure 21, showing stress concentrations for a single track bridge. As the edges of the cope holes are flame-cut, the author recommends fatigue assessment be based on hot spot stress analysis by FE-models and comparison with fatigue class 140 MPa.

The stiffener to deck plate welding is particularly important. During loading tests, the crossing of a single lorry caused the spectrum shown in Figure 22. This causes high fatigue damage. Such stress variations have never been found for railway loading, in spite of the poor distribution of patch loads. Further research [6] has shown, this is due to the action of the ballast, which redistributes the rail loading directly to the stiffener to deck plate connection and hence to the stiffener webs, in stead of causing transverse bending of the deck plate itself. This is further proven by FE-analysis as shown in Figure 23, where the vault action of the ballast is demonstrated.

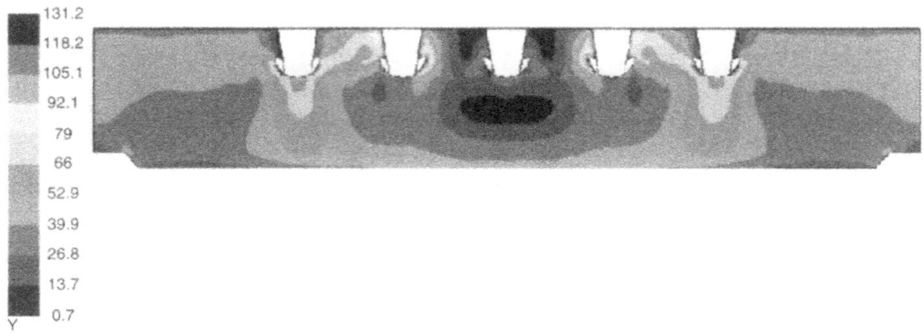

Figure 21. Stress concentrations around cope holes and cut-outs in single track bridge.

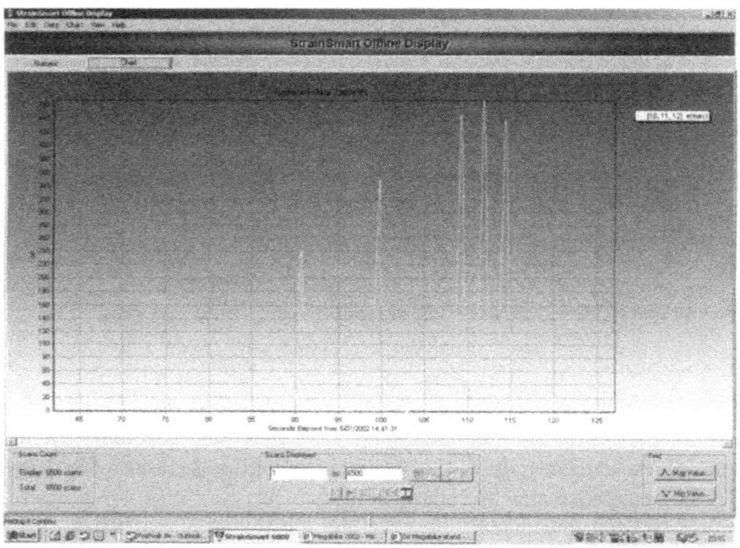

Figure 22. Measurements output of stress variations at stiffener to deck plate for crossing lorry.

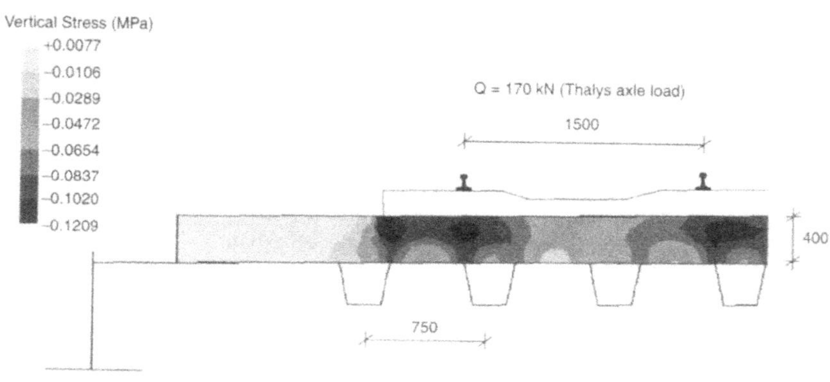

Figure 23. FE-simulation of ballast action.

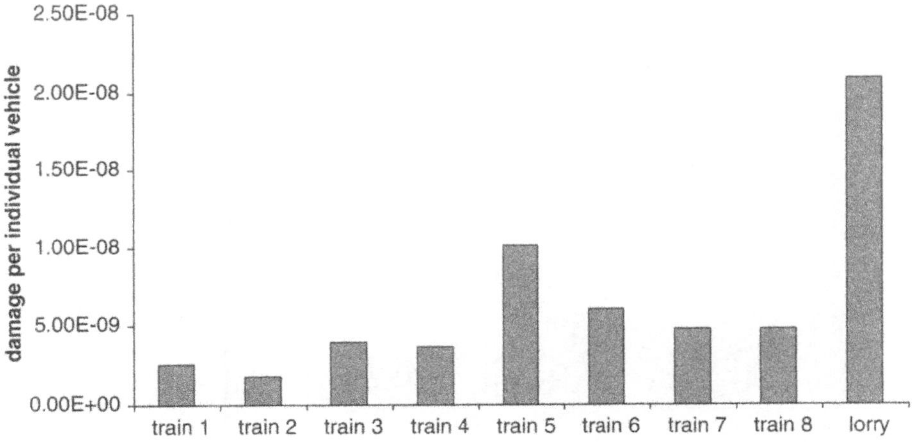

Figure 24. Individual damages caused by train vehicles and lorries stiffener to deck plate.

Comparing the fatigue damage at the stiffener to deck plate connection for each individual vehicle, as the 8 train types defined by Eurocode and the standard lorry, used in the experiments, the results of Figure 24 are obtained. This proves that steel orthotropic plated decks are a very well adapted solution for railway loading.

5 FURTHER RAILWAY EXTENSIONS

5.1 *Railway bridges in suburban areas combined with historic structures*

Although very much attention is given to the design of the large bridges which were built in the recent past, the projects for such splendid structures being numerous, designers are also requested to find original structural solutions for smaller bridges, located in surroundings which show historical value. Society has become very much aware of the possibilities of modern bridge design and is requesting more adequate proposals for these problems. Proposing original and valuable solutions for smaller bridges, located near to structures and buildings from the past, has become a challenging occupation for bridge designers.

Building next to a historic structure may need to set up a series of criteria [7], which may be used during the design process. A possible general set of criteria might be:

- the new structure should allow to keep a clear view on the most valuable parts of the historic structure;
- if the old structure has a characteristic static system (for instance groups of spans, clamped edges) it may be considered to give the new bridge an identical system;
- in addition to the former, the span length and arrangement may be chosen equal or preferably larger than for the initial structure;
- the texture of the material of the new structure will need to be adapted to the old structure's material;
- building something identical as the existing construction will never work and adversely it might be better to show the contrast between the old structure and the capabilities of modern technology;
- however, if we want to preserve and value the structures from the past, the new structure cannot be imposing or eye-catching and a delicate balance must be obtained between the two concepts.

Several cases, the writer had to design, were concerned with placing additional tracks along existing railway lines, mostly in suburban environment. A series concerned building new bridges

Figure 25. Steel trough bridge next to brickwork vaulted arches.

next to brickwork vaulted arches. A bridge across the river Senne in Brussels is located in an industrial area, which recently was renovated. The industry is of a mild type, without heavy production units. The bridge itself has a remarkable intrados curve as it slightly moves inward near the arch springs. The circular intrados is identical for the three vaults, the two approach spans fading into the earth backfill. In addition the brickwork does not show any cracking as do most of these bridges. Building a steel arch or pure concrete beam deck, would probably hide too much of this existing structure. In addition, the basic idea of the disappearing arches into the earth backfill clearly gives a hint for using a similar supporting system. Clearly, the depth of the new adjacent bridge decks must be minimised in order to keep as much as possible of the view on the vault and obtaining a slender structure. Again, steel would be a better material choice for combining with the brickwork. Hence, a central span consisting of two lateral steel trough bridges of 30 m span (Fig. 25) are supported by concrete tetrapode structures, with a concrete top slab. A particular arrangement is made for the lateral steel girders, since these have a trapezoidal cross-section. This allows provisions for diaphragm stiffeners at the intersections with crossbeams. In addition, the trapezium section decreases the slenderness of the beam webs. The width of the new decks is sufficient to contain also walkways for personnel.

When finished, the structure should appear is in Figure 25, which complies completely to the requirements as defined previously. A small part of the side vaults is covered by the new concrete supports and the deck slenderness shows an interesting contrast with the massive old structure.

In the similar case of two bridges near the town of Mechelen, brickwork vaulted arches have to be widened for additional railway tracks. This brickwork vault is supported by high piers and is remarkable due to its height. At some distance a second arch can be seen through the opening of the one in the front. The new tracks will be built at a lower level than the existing railway. This is a particular condition, requiring particular attention in the design of the bridge. Inevitably, the new structure will hide the major parts of the brickwork vault.

A steel superstructure has been considered. Even when the bridge deck reaches from the earth backfill top to top, the span is limited to 28 m. This is a very short span for a steel bridge and the most simple types of steel structure, like lateral girders or smaller trusses, should be considered. Both types have a rather bad image as they are identified to industrial sites and to 19th century construction. After consideration of this, a combination of variable depth plain I-girders and a truss

Figure 26. Hybrid I-girder and truss steel superstructure of 28 m span.

Figure 27. Tubular arch at Merxemse Street Antwerp.

was proposed as shown in Figure 26. The structure is certainly unusual and It may take some time to get familiarised with its shape, although it is immediately identified as a railway bridge. However, after evaluation by many persons it was regarded as the best of all proposals.

On the high-speed section from Antwerp to the Netherlands a local street had to be crossed, where existing tracks are supported by brickwork vaults. After considering several alternatives, a tubular steel arch bridge was chosen as shown in Figure 27. Increasing the span to 27.4 m allowed seeing the vaults through the structure. The single tube has a diameter of 0.914 m and the top concrete bridge deck is supported by 32 steel tube struts of 457 mm diameter, connected to the central tube. The struts are inclined to constitute the internal members of a truss, the lower tube and the upper concrete deck being the truss edge members. The design required particular attention for the tube nodes and their fatigue resistance. Hence the use of partial penetration welding appeared necessary, which was carried out with much care. In addition, the upper nodes, connecting steel tubes to a concrete deck by means of stud connectors required special detailing. The arch springs are determined by the torsion moments due to unsymmetrical live loads.

Figure 28. Steel tied arch bridge with radial converging hangers on high-speed line.

The Merxemstreet Bridge has also been extensively tested. At this time, the tracks were not yet installed on the bridge deck. Hence, heavy lorries were used as test load. The results of the test are reported in [8]. The structure shows large influence from integral action, since the concrete top slab is fully fixed to both abutments.

5.2 *Developments for medium span bridges*

To the North of Antwerp the high-speed line towards the Netherlands joins the motorway E 19 and crosses a number of larger roads. This required building 2 bridges with a span of 60 m. A particular arrangement with hangers converging radial to a centre located outside and below the structure's perimeter was developed. The concept has been evaluated and compared to the vertical hanger arrangement [9]. In principle, the radial converging hanger concept does perform worse than vertical hangers or certainly compared to triangular hanger arrangement. However, the variations of stresses with traffic load in the hangers and tie chord of the radial arrangement are not significantly different from the vertical hanger arrangement. Hence, if fatigue resistance is becoming an important design criterion, as is often the case for medium size railway bridges, the converging radial hanger arrangement may be an interesting alternative. Figure 28 shows a view from the air of one of these bridges.

The idea of tubular arches seemed also attractive in the case of the crossing of a double road by railway tracks on different levels. The new tracks will be used by trains running from the Brussels airport towards the cities of Antwerp and Brussels. This new extension is programmed in the years to come. Since the area below the bridge is completely taken by sewers and cable facilities, preferably no piers were to be used. The condition of having two lateral tracks at high level and an intermediate lower track, enabled to develop the idea of building 2 arches with upper concrete decks, the lower central deck being suspended to the arches. The span of this bridge reaches 78 m and the track alignment showed the difficulty of having slight bends as well as unparallel upper level tracks.

6 CONCLUSIONS

Throughout the former discussion the awareness of the effects of accelerations and comfort conditions due to high-speed railway traffic has been shown in various bridge designs. However, the effects of high-speed as well as fatigue damage caused by numerous cyclic loading from train cars can easily be overcome, by providing sufficient stiffness of bridge superstructures and develop accurate detailing of steel structures. In addition, the positive action of ballast may be stressed.

Figure 29. Twin tubular arch carrying 3 tracks, the central being suspended to the lateral arches.

Both effects will become clear also on the North and East branch of the Belgian high-speed network as soon as these sections will be commissioned. Additional measurements, mainly of structural accelerations, are to be carried out on the Prester Bridge and probably on 4 concrete viaducts.

Although railway bridges will always be rather heavy structures, a considerable effort can be done for improving the aesthetics of bridge designs and to build structures not only lasting but also agreeable to society.

REFERENCES

[1] European Committee for Standardization. 2002. *EN 1990: Eurocode Basis for Structural Design Annex A2 Application for Bridges*. CEN, Nov., 29 p.

[2] Van Bogaert, Ph. 1993. Dynamic Response of Trains Crossing Large Span Double-Track Bridges in *Journal for Constructional Steel Research*. Ed. P. Dowling and J. Harding, Elsevier Applied Science Publishers, p. 57–74.

[3] Van Bogaert, Ph. 1997. Deformation Compatibility of Steel Truss and High-Strength Concrete Girders, in *Proc Int Conference on Composite Construction – Conventional and Innovative*. Ed IABSE, CEB, CIB, ECCS, FIP, RILEM and ASCCS., Innsbrück, p. 892–893.

[4] Van Bogaert, Ph. 2003. Steel Tied Arch Railway Bridge, Louvain, Belgium, in. *Structural Engineering International – Journal of IABSE*, Vol 13 Nb 1 February, Zürich p 22–24.

[5] Raemdonk, W. 1999. Deformation Requirements of Railway Viaducts for High-Speed- Confrontation with Results from Tests, (in Dutch), *Graduation Thesis,* Ghent University, 160 p.

[6] De Corte, W. 2005. Fatigue Behaviour of Steel Orthotropic Bridge Decks subjected to Traffic Loads, (in Dutch), *Doctoral Thesis*, Ghent University, 486 p.

[7] De Pauw, B. and Van Bogaert, Ph. 2003. Widening of a historic multiple-arch concrete viaduct by integrated new fly-overs, in *Proc 2nd Int. Speciality Conference on The Conceptual Approach to Structural Design*, Ed. F. Mola, Milan, July, pp. 363–369.
[8] Van Bogaert, Ph. 2006. 'Design and Construction of the Merxem Bridge – A Single Tubular Arch'. *Proc. 6th Int. Symposium Steel Bridges.CAOK-CVUT-European Convention for Structural Steelwork ECCS.* May 31–June 2 Prague. Ed. Studnicka, J. ISBN 80-239-7168-9. pp 56–65
[9] Van Bogaert, Ph. and De Pauw, B. 2003. Tied Arch Bridges with Radial Oriented Hangers, subjected to Rail Traffic, in *Proc IABSE Symposium Structures for High-Speed Railway Transportation*. Ed IABSE, Antwerp August, CDRom and Report p. 142–143.

CHAPTER 5

Last developments of dynamic analysis of bridges under moving train loads

S. Montens & J. Huyard
SYSTRA, Paris, France

ABSTRACT: Dynamic calculation of bridges under moving train loads have been used for more than 20 years. The first analysis methods did not take into account, neither the interaction between the bridge and the rolling stock, nor the transverse behaviour, nor the track imperfections. We developed recently all these new aspects. The true interaction between the vehicles and the bridge can been taken into account, both for vertical and horizontal movements. The track imperfections can be taken into account also. The following results can be obtained: bridge deflection and acceleration, vehicle acceleration, minimum wheel load, derailment coefficient.

1 DESCRIPTION OF TRADITIONAL DYNAMIC ANALYSIS UNDER MOVING TRAIN LOADS

Following the design of the first high speed train line in France, dynamic calculation of bridges under moving train loads has been developed ([1]).

The purpose of this analysis was triple:

- check the train running safety, both concerning the ballast stability and the rail-wheel contact,
- check the bridge strength and fatigue,
- check the passenger's comfort.

The problem was simplified in the following way. First, a dynamic analysis of the train was performed, considering that the train applies forces to the bridge deck, which are constant with time: these forces move, but their intensity remains the same. This is equivalent to separating the bridge dynamic behaviour, and the vehicle dynamic behaviour, assuming that they are uncoupled.

The calculation scheme was then the following:

- modal analysis of the bridge structure,
- search of critical speeds at which the resonance phenomena are possible,
- calculation, at these critical speeds as well as specific speeds on the line, of: the maximum vertical deflection of each track in each span, the maximum vertical accelerations at each track axis, the rotations of bearing sections at each deck end, and the maximum track twist.

Then, the dynamic analysis of the vehicle is performed. A simplified dynamic model of the vehicle is built, including the wheels, the bogies and the vehicle body, linked by springs and dampers. The vertical displacement z(t) of a given bogie (or wheel) is extracted from the dynamic analysis of the bridge performed before. By integration over time of the conventional differential equation of the vehicle movement, the vertical displacements and accelerations of the vehicle body are calculated. The maximum vertical acceleration is checked towards the comfort criteria.

2 DYNAMIC TRANSVERSE ANALYSIS

2.1 *General*

Transverse vibrations of bridge structure under railway loads can be created by two phenomena.

Transverse vibrations can be created mainly by track imperfections. Due to track imperfections (alignment), the train wheels impact transversely the rails, and apply horizontal transverse forces to the bridge deck, which then moves transversely.

Secondly, in the common case where a deck supports two tracks, when a train is running on one track only, there is some torsion in the deck, and so some transverse bending of the pier shaft. This transverse bending creates some transverse displacement of the pier top, and of the deck, and so some transverse acceleration of the deck.

These effects are generally negligible, but they could become a problem, as it happened in some case, if the track imperfections were great and if the pier transverse stiffness was very low.

2.2 *Modelling of transverse behaviour*

In order to consider the transverse behaviour, it is possible to keep the same general methodology as described above for the vertical behaviour, assuming uncoupled behaviour of the bridge structure and of the vehicle.

Of course, it is necessary to incorporate into the bridge model the characteristics of the piers and foundations, which was not the case for the vertical behaviour.

The dynamic model of the vehicle must incorporate the wheels running on both rails.

3 TRACK IMPERFECTIONS

3.1 *General*

The track geometry is never perfect. There are some imperfections regarding: the level of each rail, and the transverse layout of each rail.

These imperfections can be quantified by their power spectral densities, which correspond to the magnitude of the defects according to their length.

The American rules, for example, specify 5 categories, from the best to the worst quality of geometry.

The actual track imperfections can also be measured on a given line, and converted into power spectral densities (see Fig. 3).

3.2 *Calculation scheme*

3.3 *Generation of track irregularity*

Calculations are performed in the time domain. So it is necessary first to generate time-history of track irregularity from the power spectral density of track irregularity.

For a given train speed, trajectory of track irregularity is calculated as follows:

$$t_a(x) = \Sigma_i A_i cos(\omega_i t + \varphi_i) \quad (1)$$

$$A_i = (2.Sa(\omega_i) \cdot \Delta\omega)0.5 \quad (2)$$

$t_a(x)$: trajectory of the alignment irregularity
$Sa(\omega)$: power spectral density of alignment irregularity
ω: spatial angular frequency
φ_i: randomly chosen phase

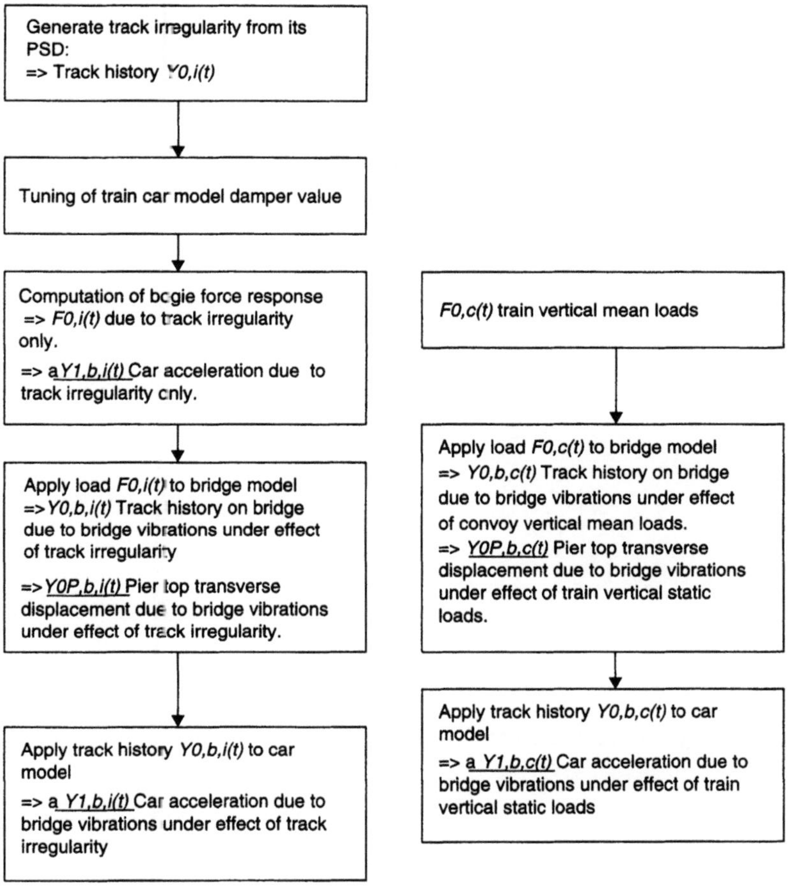

Figure 1. Calculation scheme with track imperfections.

The following is an example of track irregularity trajectory for a freight train line of poor quality, for a given train speed (55 kph).

3.4 *Train car models*

3.4.1 *Freight train car model*

Some freight train cars have a non linear behavior, moreover damping is due to friction. The following assumptions can be adopted in order to simplify its model:

Connection between rail and wheel-set supports is rigid (conservative assumption)
Damping is modeled by an equivalent viscous damper
Car mass is lumped at its extremities, i.e. on bogie

$Y0$: transverse track trajectory input of the model. This trajectory is the mean trajectory of the bogie, i.e. the mean of the bogie two wheel-sets trajectories.
$Y1$: the response transverse displacement of the car center of mass
PHI: the response longitudinal axis rotation of the car center of mass
Cz is tuned in order to dissipate the same energy quantity than the friction damper along the same loading history.

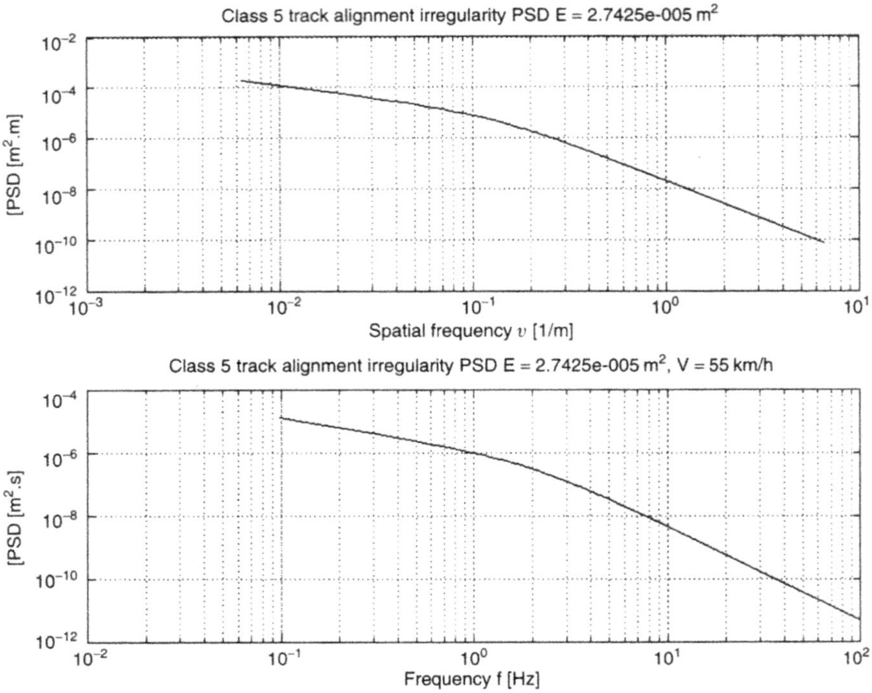

Figure 2. Track irregularity spectrum for freight train.

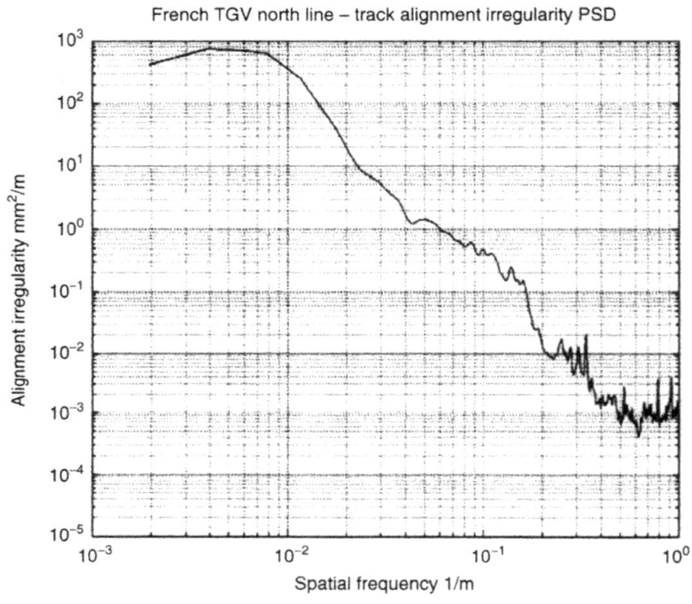

Figure 3. Track irregularity spectrum for high speed train.

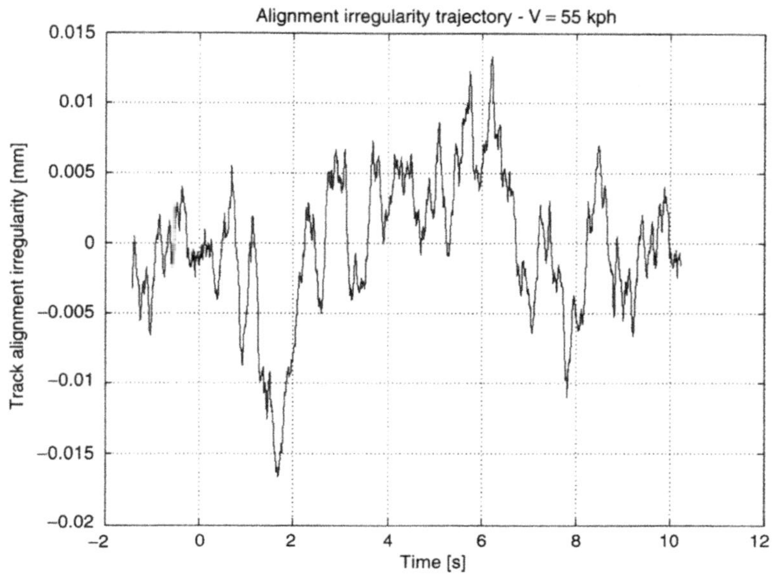

Figure 4. Track irregularity trajectory for a freight train line.

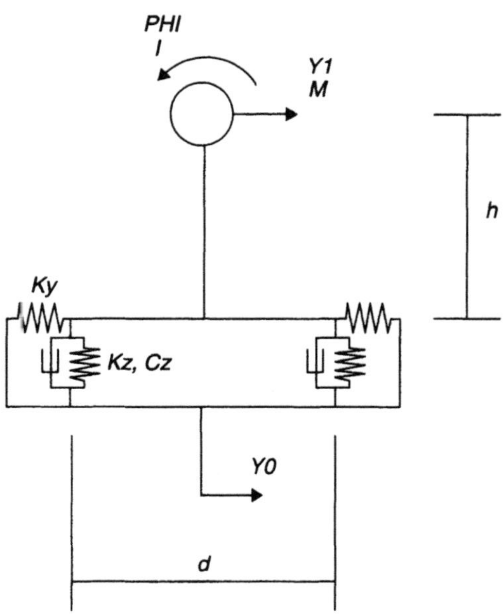

Figure 5. Freight train car transverse model.

Motion equations are implemented in Matlab as a state space model, then 2 digital filters are derived:

- $Y0$ as the input and $Y1$ as the output.
- $Y0$ as the input and PHI as the output.

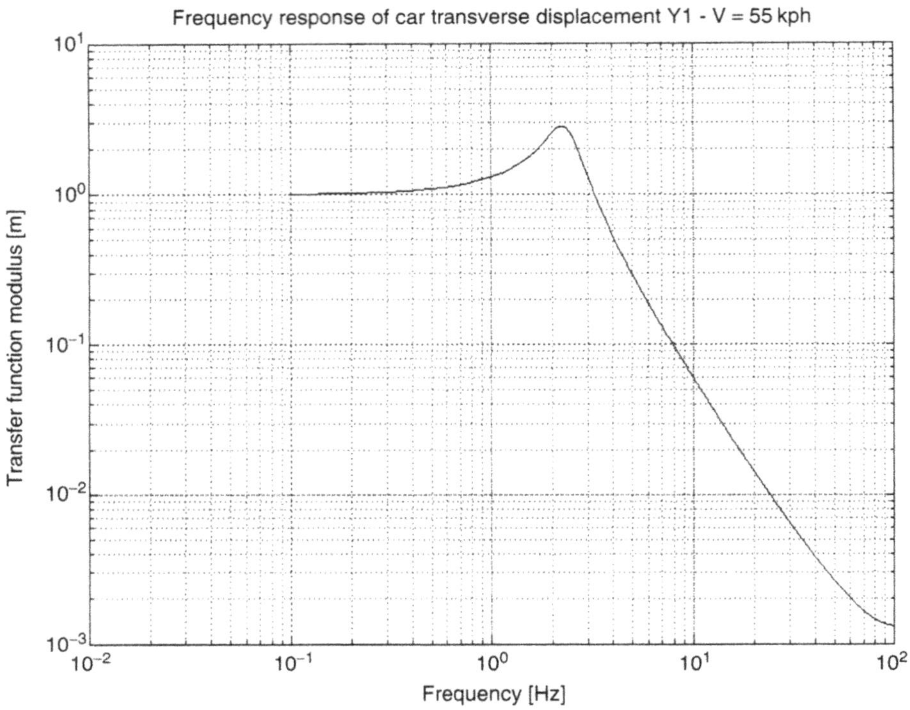

Figure 6. Transfer function modulus of freight train car.

Viscous damper value is chosen in order to dissipate the same energy than friction damper along the same load path.

The transfer function modulus of car transverse displacement is given below:

We assumed a freight type train composed with 30 cars.

3.4.2 *High speed train car model*

Characteristics of the TGV, French high speed train, are used. Nominal loads and spacing between loads are shown hereafter,

Y0: transverse track trajectory input of the model. This trajectory is the mean trajectory of the bogie, i.e. the mean of the bogie 2 wheel-sets trajectories.

These wheel-sets are 3.00 m spaced longitudinally.
Y1: the response transverse displacement of the bogie center of mass
Y2: the response transverse displacement of the car center of mass
PHI1: the response longitudinal axis rotation of the bogie center of mass
PHI2: the response longitudinal axis rotation of the car center of mass

The transfer function of car transverse displacement is given below (modulus and argument):

3.5 *Effects of track irregularity outside the bridge*

The maximum train car transverse acceleration is 1.7 m/s^2.

The maximum train car transverse acceleration is 0.18 m/s^2 and the mean frequency is 5.30 Hz at V = 255 kph.

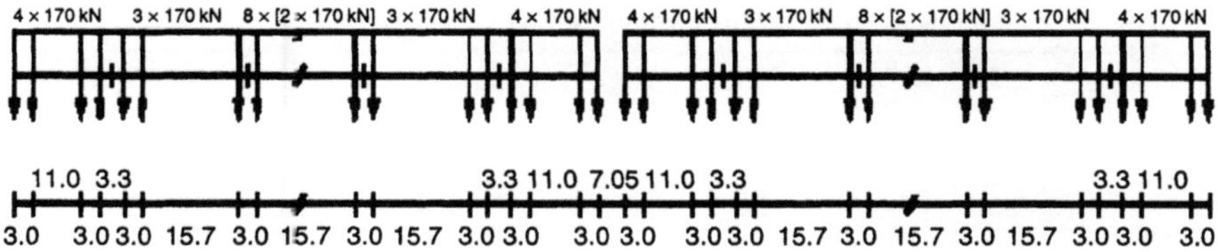

Figure 7. TGV Nominal loads.

TGV train is composed with following car sequence:
[M + R1 + 8 × R2-R9 + R10 + M] + [M + R1 + 8 × R2-R9 + R10 + M]
First train Second train

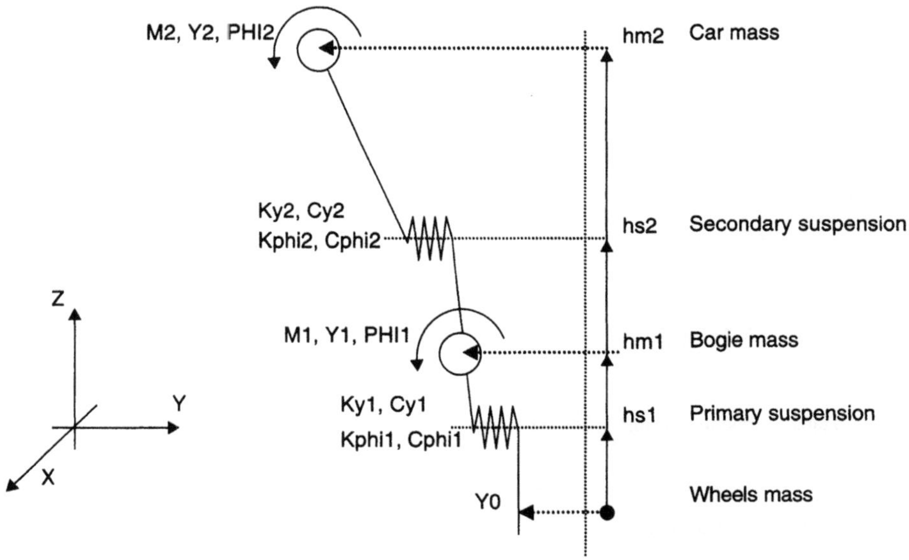

Figure 8. High speed train car transverse model.

3.6 *Bridge model*

The bridge is modeled with ANSYS software. Pier and decks are modeled with BEAM4: 3D beam element; additional SIDL mass with MASS21 element: 3D mass with rotation inertia. Foundations are represented by an equivalent stiffness with COMBIN14 spring elements.

Full transient analysis is then carried out in order to estimate the response of the bridge:

1. Under vertical train static loads,
2. Under previously computed bogie force induced by track alignment irregularity.

These calculations give:

- transverse displacement time-history of the pier top
- bogie displacement time-history on the bridge

3.7 *Pier top transverse displacement*

The main results are the following:

On the first graph, we observe, under vertical train nominal (or static) loads, a quasi-static transverse displacement; its value is around 0.9 mm. The maximal total transverse displacement is 3.00 mm. So the maximal dynamic transverse displacement is 2.1 mm.

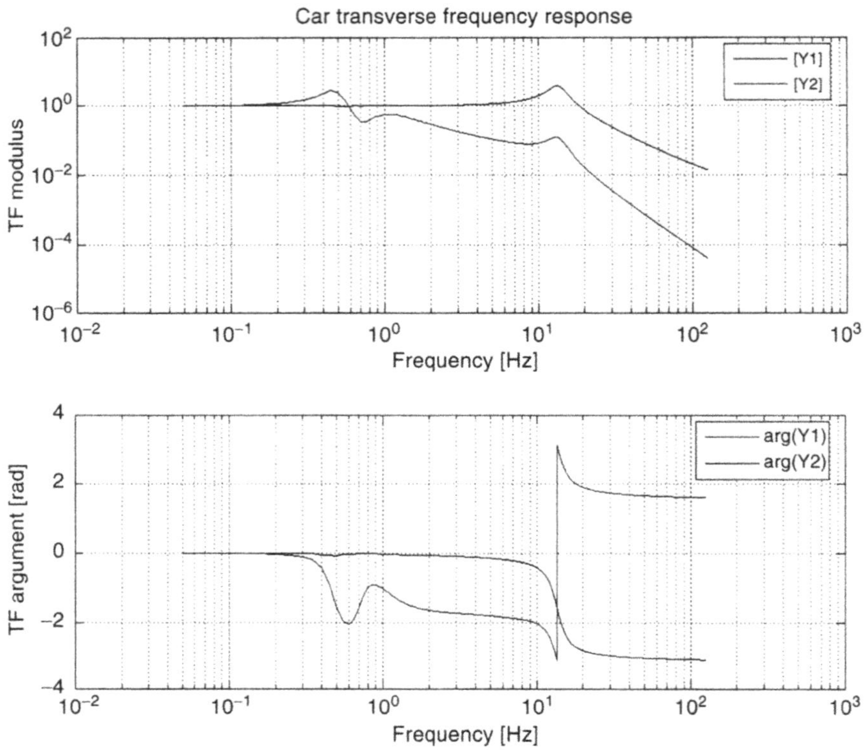

Figure 9. Transfer function of high speed train car.

Table 1. Pier top transverse displacement.

Train type	Freight train	High speed train
Maximum displacement	3 mm	1.1 mm
Speed	55 kph	95 kph
Dynamic displacement	2.1 mm	0.75 mm

On the first graph, we observe, under vertical train nominal (or static) loads, a quasi-static transverse displacement; its value is around 0.35 mm. The transverse displacement under transverse load is negligible. The maximal total transverse displacement is 1.10 mm.

So, the maximal dynamic transverse displacement is 0.75 mm.

3.8 *Car transverse acceleration due to bridge vibration*

For the freight train we chose to observe car #27 in order to take into account maximal bridge movement induced by preceding cars. Car maximal acceleration is obtained for V = 55 kph. It is equal to 0.57 m/s^2.

For the high speed train we chose to observe car #10 of the second train in order to take into account maximal bridge movement induced by preceding cars. Car maximal acceleration is obtained for V = 110 kph. It is equal to 0.013 m/s^2.

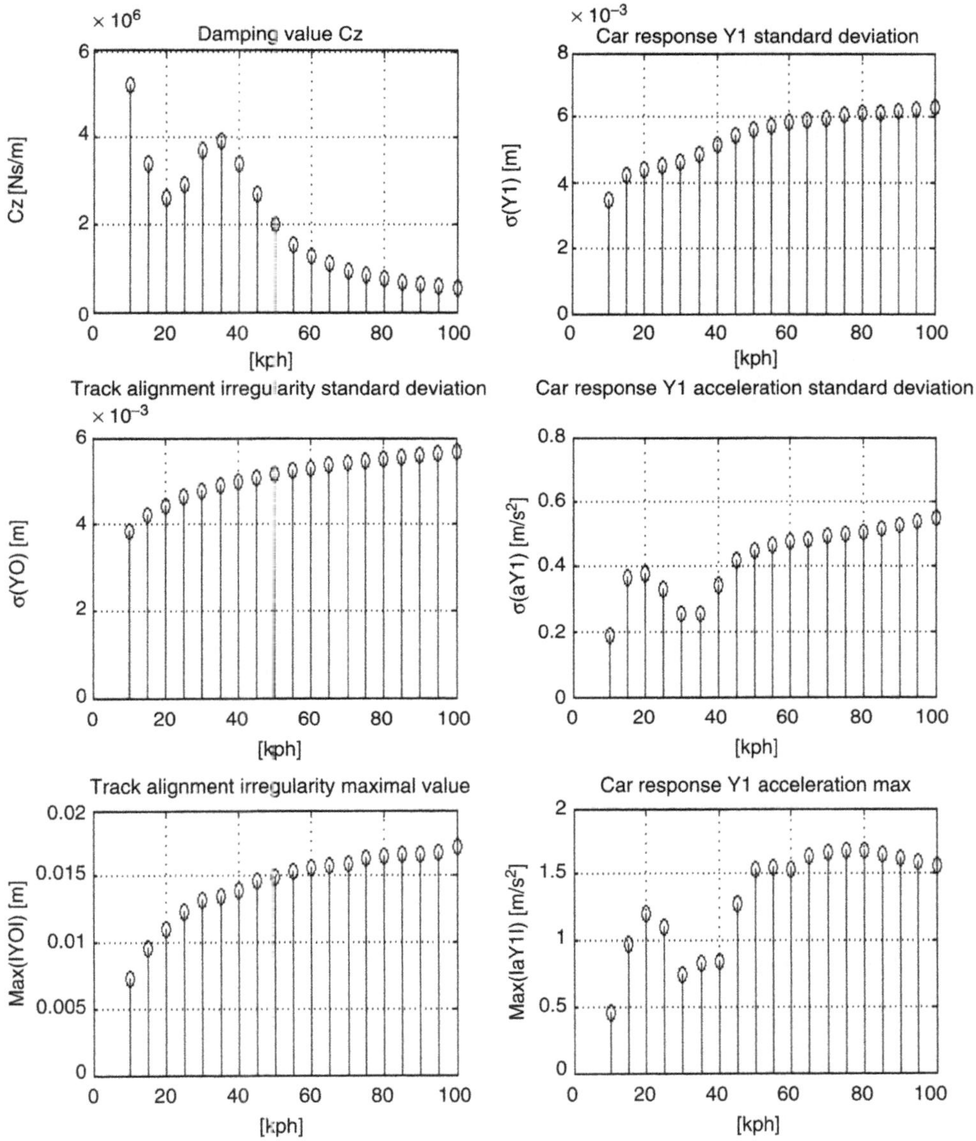

Figure 10. Effect of track irregularity outside the bridge for freight train.

3.9 *Freight train car transverse acceleration final results*

Freight train car peak acceleration:

To obtain car peak transverse acceleration while running on the bridge, we compose both results for V = 55 kph:
Outside the bridge: ay_max_1 = 1.55 m/s^2 (mean vibration frequency 3.1 Hz)
Due to the bridge vibrations: ay_max_2 = 0.57 m/s^2
ay_max = (ay_max_1^2 + ay_max_2^2)^0.5
On the bridge, ay_max = 1.65 m/s^2

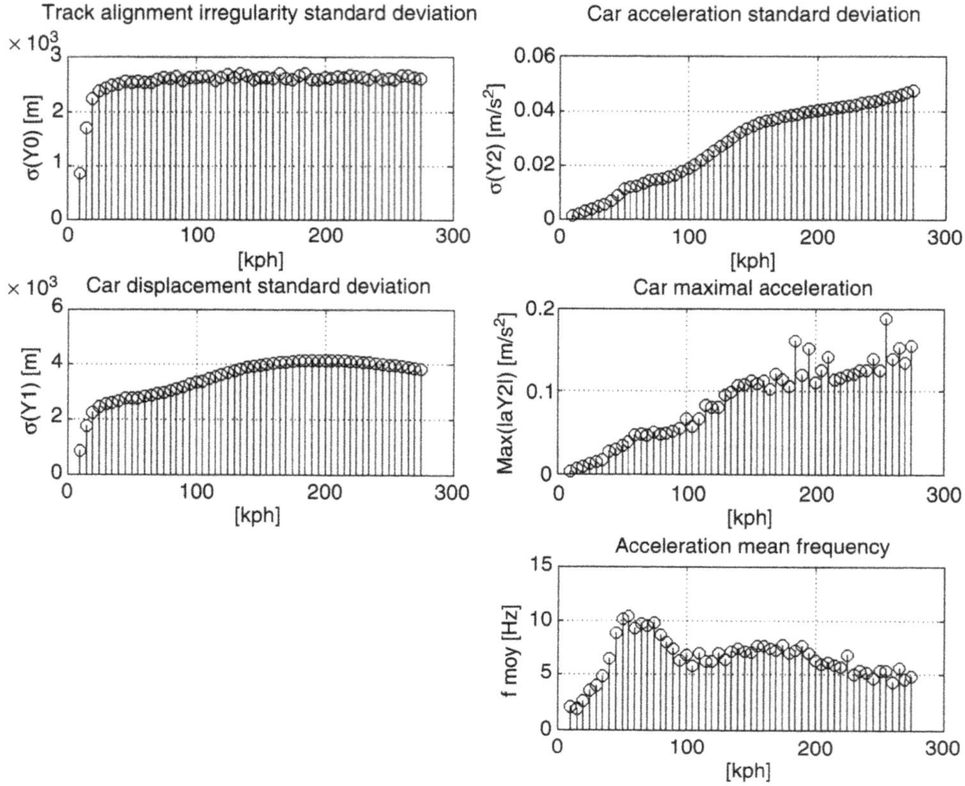

Figure 11. Effect of track irregularity outside the bridge for high speed train.

Outside the bridge, ay_max = 1.55 m/s^2
The acceleration increase due to the bridge is only 6%.

3.10 *High speed train car transverse acceleration final results*

High speed train car peak acceleration:

To obtain car peak transverse acceleration while running on the bridge, we compose both results for V = 255 kph:
Outside the bridge: ay_max_1 = 0.180 m/s^2 (mean vibration frequency 5.30 Hz)
Due to the bridge vibrations: ay_max_2 = 0.013 m/s^2
ay_max = (ay_max_1^2 + ay_max_2^2)^0.5
On the bridge, ay_max = 0.18 m/s^2
Outside the bridge, ay_max = 0.18 m/s^2
The acceleration increase due to the bridge is negligible.

4 COMPLETE INTERACTION ANALYSIS

4.1 *Calculation scheme*

The calculation scheme is the following. The bridge model dynamic equations and the train car dynamic equations are solved in a complete system of equations, using Matlab software, so that there is interaction between the bridge and the train cars.

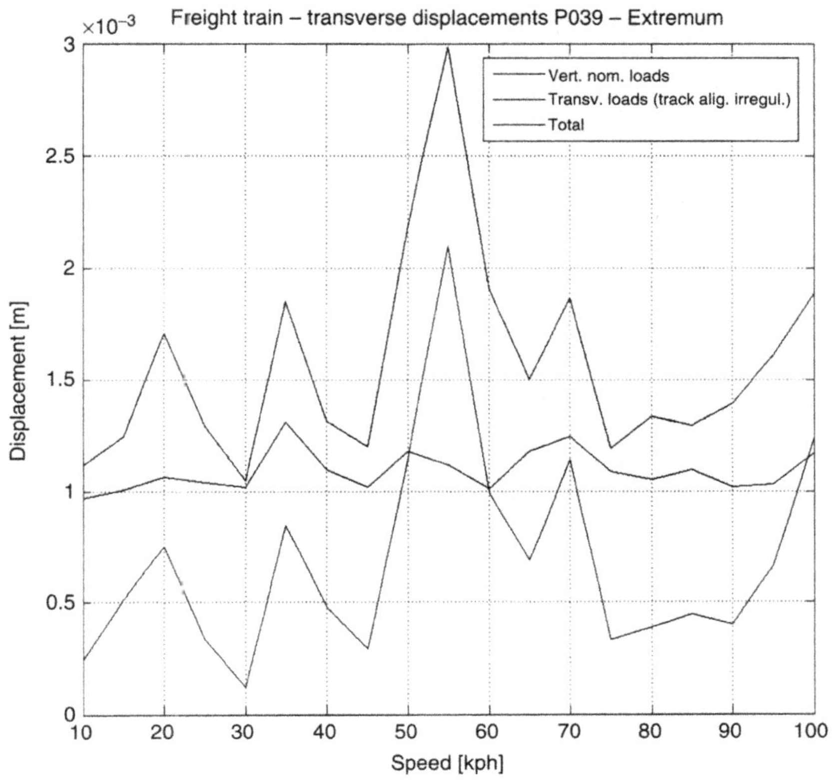

Figure 12. Pier top transverse displacement under freight train, as function of speed.

4.2 *Track irregularity*

Track irregularity PSD for Freight train

We take into account following PSD for vertical irregularity

$$S_v(\Omega) = k \cdot A_v \cdot \Omega_c^2/[\Omega^2 \cdot (\Omega^2 + \Omega_c^2)] \quad (3)$$

With $k = 0.25$
$\Omega_c = 0.8245$ rad/m
$A_v = 0.2095e - 4\,m^2 \cdot$ m/rad
taken from American class 5 power spectral density.
Track irregularity PSD for TGV train
Track irregularity for TGV train is issued from measures on TGV French North line.

4.3 *Train car vertical model*

TGV train car vertical model

$Z0$: vertical track trajectory input of the model. This trajectory is the mean trajectory of the bogie, i.e. the mean of the wheelsets trajectories. These wheel-sets are 3.00 m spaced longitudinally.
$Z1$: the response vertical displacement of the bogie center of mass
$Z2$: the response vertical displacement of the car center of mass

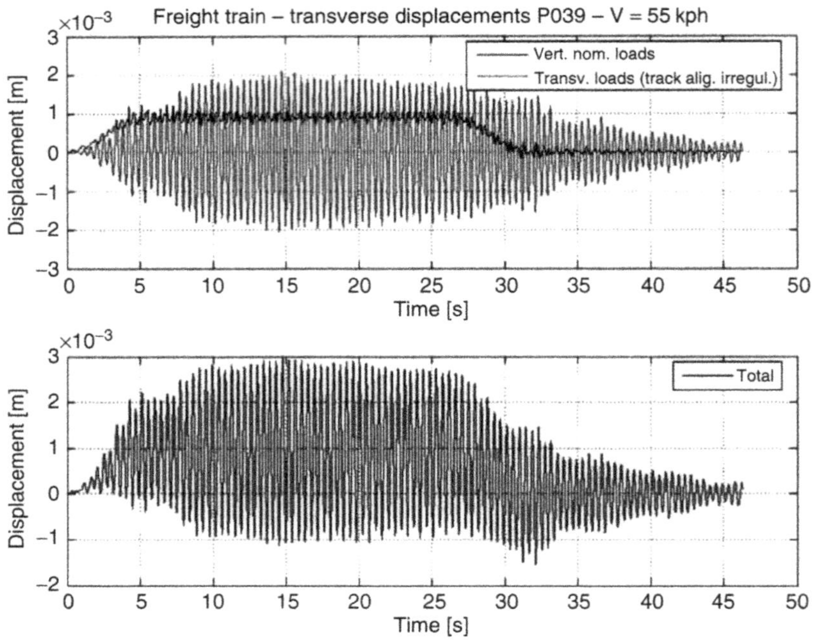

Figure 13. Pier top transverse displacement under freight train for V = 55 kph.

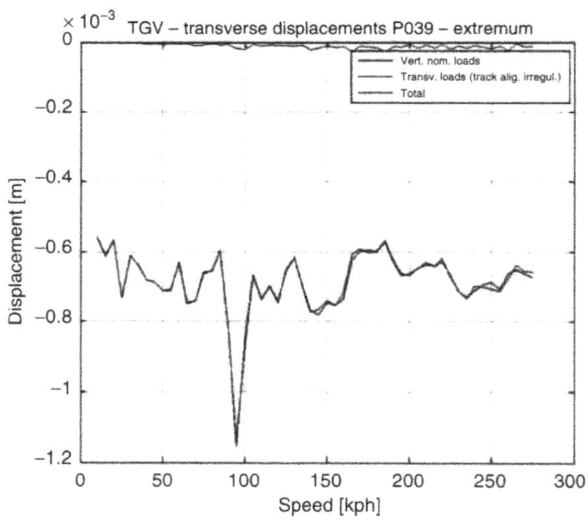

Figure 14. Pier top transverse displacement under freight train as function of speed.

Motion equations

$$M1d^2/dt(Z1) = -K1(Z1 - Z0) - C1d/dt(Z1 - Z0) - M2d^2/dt(Z2) \tag{4}$$

$$M2d^2/dt(Z2) = -K2(Z2 - Z1) - C2d/dt(Z2 - Z1) \tag{5}$$

These equations are implemented in Simulink as a state space model, with *Z0* as input and *Z1* and *Z2* as outputs:

$$d/dt(\mathbf{Z}) = A \cdot \mathbf{Z} + B \cdot \mathbf{U} \tag{6}$$

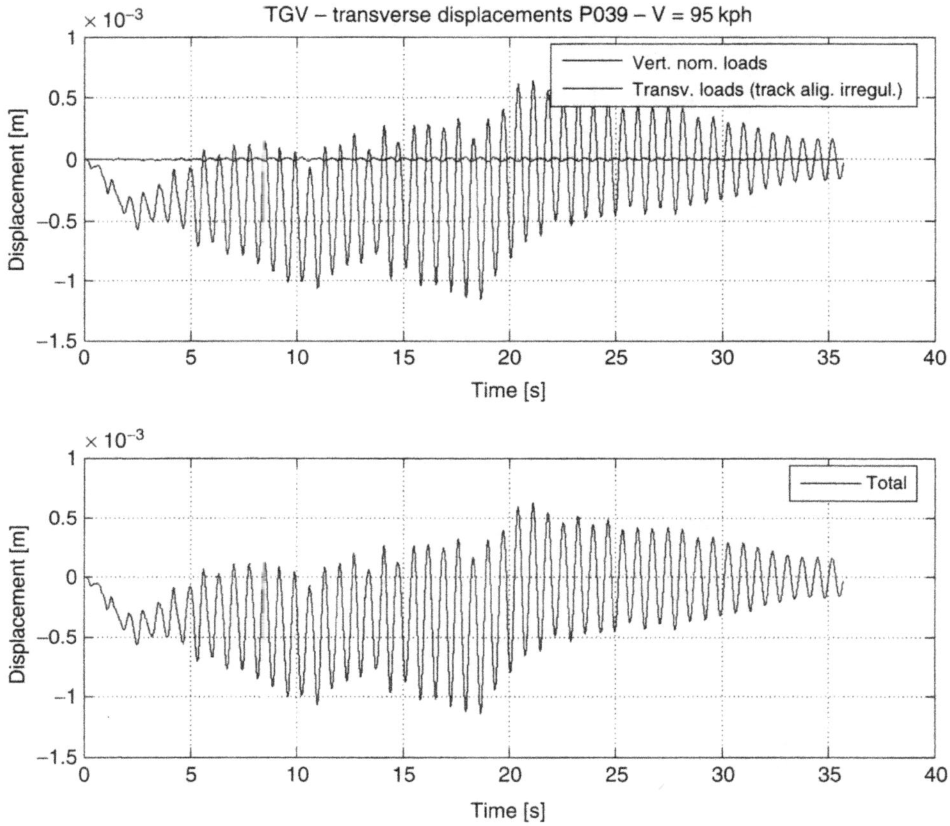

Figure 15. Pier top transverse displacement under high speed train for V = 95 kph.

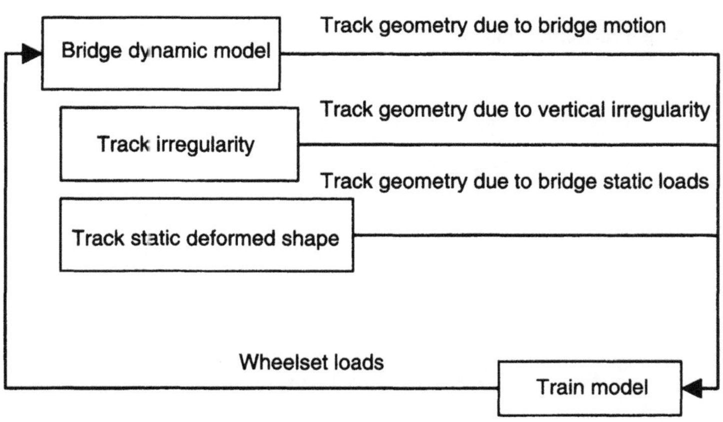

Figure 16. Calculation scheme for vertical dynamic analysis.

Freight Train car vertical model
Motion equations

$$Zc = 1/2 \times (Zc, 1 + Zc, 2) \tag{7}$$

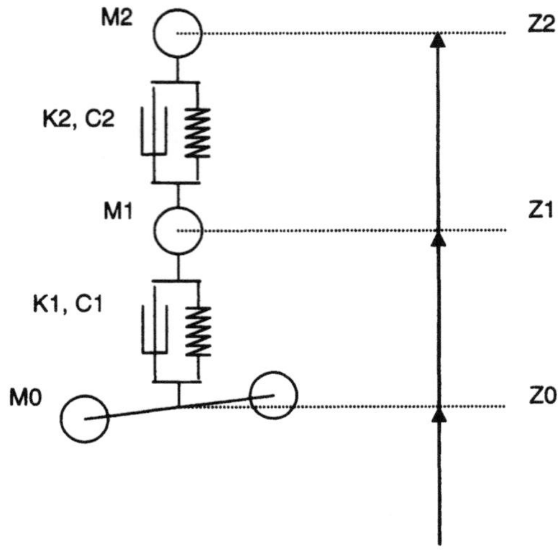

Figure 17. TGV train car vertical model.

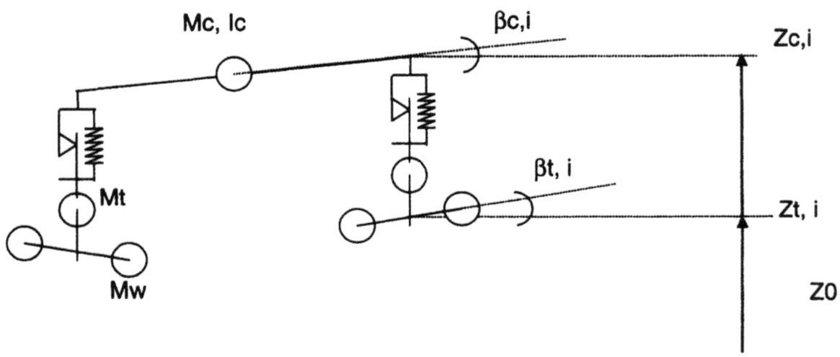

Figure 18. Freight train car vertical model.

$$\beta c = 1/(2 \times 4.25) \times (-Zc, 1 + Zc, 2) \tag{8}$$

$$Fz, i = Ktz \cdot (Zc, i - Zt, i + dZ0) + f \cdot Pa \tag{9}$$

$$f = f0 \quad \text{if d/dt}\,(Zc, i - Zt, i) > 0 \tag{10}$$

$$f = f1 \quad \text{if d/dt}\,(Zc, i - Zt, i) < 0 \tag{11}$$

$$My, i = \mu \cdot N \cdot lrx \cdot sg[d/dt(\beta c, i - \beta t, i)] \tag{12}$$

$$N = Nl \quad \text{if d/dt}\,(Zc, i - Zt, i) > 0 \tag{13}$$

$$N = Nu \quad \text{if d/dt}\,(Zc, i - Zt, i) < 0 \tag{14}$$

$$Pa = KZ1 \cdot (Zc, i - Zt, i) \tag{15}$$

$$Mc \cdot \mathrm{d}^2/\mathrm{dt}^2(Zc) = -2Fz, 1 - 2Fz, 2 + Mc \cdot g \tag{16}$$

$$Ic \cdot \mathrm{d}^2/\mathrm{dt}^2(\beta c) = (2Fz, 1 - 2Fz, 2) \cdot lc - My, \tag{17}$$

These motion equations are directly implemented in Simulink, inputs are Zt,i and ouputs are Zc,i & dynamic loads applied to wheelsets.

4.4 *Bridge model*

Bridge is modeled with ANSYS software as for transverse analysis. The bridge model is the succession of 8 spans, with 43 nodes by span. For each train type, 2 analysis are carried out:

- a static analysis, in order to estimate the track 1 static deformed shape due to a train on track 2,
- a modal analysis.

Eigen values and eigen vectors of track vertical displacements are derived from modal analysis. They are implemented in Simulink as a state-space model, allowing to compute the bridge dynamic response by modal superposition method.

Other parameters are:

- damping coefficient for bridge modal superposition ksi = 0.01
- trains composition as follows:
  - 32 wheelsets (8 cars) for Freight train
  - 30 wheelsets (12 cars) for TGV
- range of studied speeds: 10–100 kph for freight train, and 100–275 kph for TGV, with 5 kph step.

4.5 *Results for TGV train*

Bridge vibrations

Maximal bridge vertical displacement 4.20 mm occurs at speed 275 kph. This value includes the track static displacement due to the trains load on both tracks (1.58 mm).

=> Bridge maximal dynamic displacement = 2.62 mm

According to Eurocode, for one track loaded, the deflection

$f = 0.00158/2 + 0.00262 = 0.0034\,\text{m} < L/1750 = 0.023\,\text{m}$

We notice a peak in bridge response for the speed around 275 kph. Spacing between consecutive cars center of mass is 18.70 m which at speed 275 kph corresponds to f0 = 4.08 Hz fundamental frequency. This excitation frequency matches with bridge first longitudinal bending mode of the TGV train deck (~4.10 Hz).

Maximal bridge acceleration 1.58 m/s^2 occurs at speed 275 kph.

=> The bridge fulfills the criteria of ballast limit acceleration $1.58 < 3.5\,\text{m/s}^2$ (Eurocode).

Car vibrations

Extreme accelerations are mainly due to track imperfections.

The maximal acceleration is $0.5\,\text{m/s}^2 < 1.0\,\text{m/s}^2$, so the comfort is very good according to Eurocode.

Maximal wheel dynamic vertical load due to vertical vibrations is $\Delta P1 = 14.5$ kN.

According to the transverse dynamic analysis, the maximal car transverse acceleration was $0.18\,\text{m/s}^2$.

Therefore, the corresponding transverse force Q, for a wheelset, is $0.5 \times 0.18 \times 33549 \sim 3.00$ kN, and maximal wheel dynamic vertical load due to transverse vibrations is $\Delta P2 \sim 3.00 \times 1.62/1.50 = 3.25$ kN

Total wheel dynamic vertical load is then

$\Delta P = (14.5^2 + 3.25^2)\hat{}0.5 = 14.9$ kN

The variation of wheel load is

$\Delta P/\text{Pmean} = 14.9/82.3 = 0.18 < 0.60$ OK

The derailment coefficient is

$Q/\text{Pmini} = 1.3 \times 3.00/(82.3 - 14.9) = 0.06 < 1.0$ OK

Q is factored by 1.3 in order to take into account the distribution (+1.3/−0.3) of the transverse force from the wheelset to the wheels.

4.6 *Results for Freight train*

Bridge vibrations

Maximal bridge displacement 7.15 mm occurs at speed 85 kph. This value includes the track static displacement due to the trains load on both tracks (6.10 mm).

=> Bridge maximal dynamic displacement = 1.05 mm.

According to Eurocode, for one track loaded, the deflection

$f = 0.0061/2 + 0.00105 = 0.0041\ m < L/600 = 0.066\ m$

Bridge accelerations are mainly caused by track imperfection.

Maximal bridge acceleration 0.95 m/s^2 occurs at speed 95 kph.

=> The bridge fulfills the criteria of ballast limit acceleration 0.95 < 3.5 m/s^2 (Eurocode).

Car vibrations

Extreme accelerations are mainly due to track imperfections.

Extreme accelerations criteria for freight trains is fulfilled since 4.2 m/s^2 is lower than 0.7 g = 6.87 m/s^2.

Maximal wheel dynamic vertical load due to vertical vibrations is $\Delta P1 = 50.0$ kN.

According to transverse dynamic analysis, the maximal car transverse acceleration, at 100 kph, was 1.6 m/s^2.

Therefore, the corresponding transverse force Q, for a wheelset, is $0.5 \times 1.6 \times 38970 = 31.2$ kN, and maximal wheel dynamic vertical load due to transverse vibrations is

$\Delta P2 = Q \times (1.339 + 0.03 + 0.15)/1.50 = 31.6\ kN$

Total wheel dynamic vertical load is then

$\Delta P = (50.0^2 + 31.6^2)\hat{}0.5 = 59.1\ kN$

The variation of wheel load is

$\Delta P/Pmean = 59.1/105.1 = 0.56 < 0.60$ OK

The derailment coefficient is

$Q/Pmini = 1.3 \times 31.2/(105.1 - 59.1) = 0.89 < 1.0$ OK

Q is factored by 1.3 in order to take into account the distribution (+1.3/−0.3) of the transverse force from the wheelset to the wheels.

5 CONCLUSION

It is possible to model the true interaction between the vehicles and the bridge, both for vertical and horizontal movements. The horizontal and transverse track imperfections can be taken into account also.

The following results can be obtained: bridge deflection and acceleration, vehicle acceleration, minimum wheel load, derailment coefficient.

REFERENCES

[1] GANDIL J., RAMONDENC P., The dynamic behaviour of railway bridges on high speed lines, IABSE Symposium, Leningrad, 1991.

CHAPTER 6

Assessment of the dynamic response of high-speed railway bridges in interoperable lines

R. Delgado, R. Calçada, D. Ribeiro, J.R. Pinto & H. Figueiredo
FEUP, Porto, Portugal

I. Faria
UTAD, Vila Real, Portugal

ABSTRACT: In this work, dynamic analyses of different structural types of bridges under high-speed railway loading were carried out, towards the assessment of its dynamic behaviour in interoperable lines, in which the Technical Specifications for Interoperability (TSI) are applicable. The analyses were made for the passage of real trains and the loading schemes relative to the High Speed Load Model. The responses were evaluated from the point of view of structural safety, track safety and passengers comfort, based on the criteria established in the EN1991-2 and EN1990-AnnexA2.

1 INTRODUCTION

A high-speed railway system is presently under implementation in Portugal, which will enable the connection of the country with a large European network. At the moment, many countries have already developed their own structural solutions for dealing with the effects of high-speed trains in bridges and a great number of structures have been in operation for several years. These solutions were initially designed for specific train types, thus being checked for the dynamic effects of only a small part of the actual European high-speed rail traffic. In the case of interoperable lines, as will be the future Portuguese high-speed network, the high-speed Technical Specifications for Interoperability (TSI) [1] are applicable, requiring additional checks for dynamic analysis using the High-Speed Load Model (HSLM).

In this paper the dynamic behaviour of different types of bridges is investigated, namely for a metallic bowstring arch bridge, existing in a conventional line, a prestressed concrete voided slab bridge, a prestressed concrete box girder bridge and a composite twin girder bridge, part of high-speed railway lines. Dynamic analyses of these structures were performed for both the actual high-speed real trains and the HSLM load schemes, in an attempt to infer the consequences of the application of these new models to the design of railway bridges. The response of the bridges were assessed in terms of structural safety, track safety and passengers comfort, based on the most recent criteria established in the EN1991-2 [2] and EN1990-AnnexA2 [3].

2 DYNAMIC EFFECTS ON RAILWAY BRIDGES ACCORDING TO THE EUROCODES

For speeds greater than 200 km/h, the dynamic effects tend to increase substantially, mainly as a result of the resonance effects. The definition of these effects is a complex task which requires detailed dynamic analyses. The procedures for the determination of these effects have been included in the latest version of EN1991-2 and EN1990-AnnexA2, evidencing the most recent research carried out by expert committees D214 [4] and D190 [5] of ERRI.

2.1 *Check on the requisites for dynamic analysis*

The decision on the need to perform dynamic analysis of a bridge for the assessment of the dynamic effects, is presented in EN1991-2 in the form of a flowchart (Fig. 1).

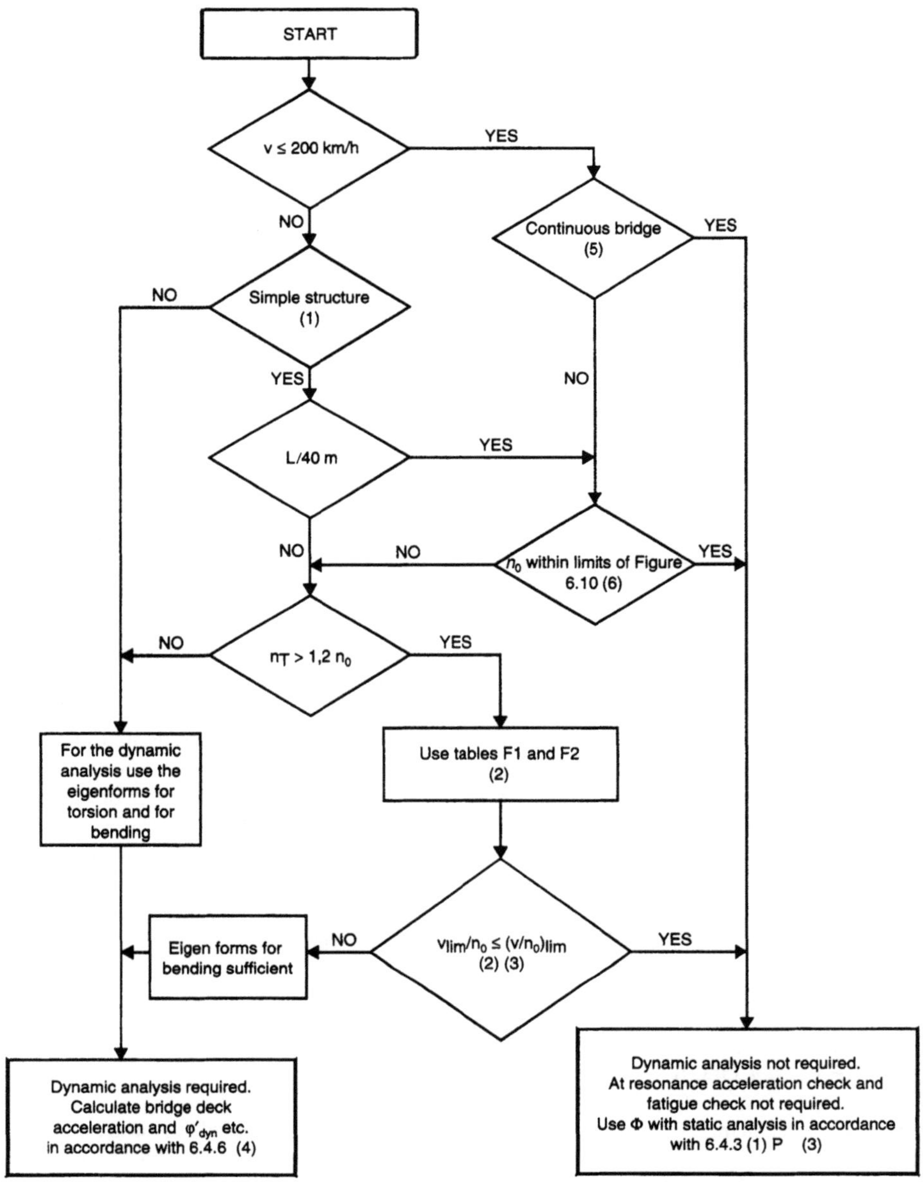

Figure 1. Flowchart for the determination of the need to perform a dynamic analysis of the bridge (adapted from [2]).

where:

v (km/h) is the maximum speed at the bridge site;
L (m) is the span of the bridge;
n_0 (Hz) is the fundamental bending frequency of the bridge;
n_t (Hz) is the fundamental torsional frequency of the bridge;
v_{lim}/n_0 (m) and $(v/n_0)_{lim}$ (m) are presented in the annex F of EN1991-2.

2.2 *Loads scenario*

A characteristic of the resonance phenomena is that its occurrence for a specific bridge depends not only on the vehicle speed but also on the type of train. In what concerns the speed, the analyses should consider speeds between 40 m/s and 1.2 times the maximum speed at the bridge site. As for the type of trains, in order to avoid circulation restrictions on the train type for a given railway, it is suggested that the analyses should consider a wide range of real trains, thus enabling variations on the exploration conditions and the possibility of connexions with other European high-speed railways.

2.2.1 *Real trains (RT)*

The real trains of the European high-speed railway network are of the following types (Fig. 2): i) Conventional, in which each passenger's carriage has two bogies, each of those with two axles. The ICE 2, ETR-Y and VIRGIN trains belong to this type; ii) Articulated, where each passenger's carriage has one bogie of two axles at each end, shared with the adjacent carriage and centered in the articulation of the two. The THALYS, EUROSTAR and TGV trains are of this type; iii) Regular, in which the carriages are also articulated, though the articulation is not supported by the bogie but on a single axle between the two carriages. The TALGO train is one of this type.

2.2.2 *High speed load model (HSLM)*

Bridges located in high-speed railways which comply with the interoperability criteria must be analysed from the point of view of their dynamic behaviour taking into account the HSLM loading models (A or B). The loading model HSLM-A consists of ten universal trains (A1 to A10), as schematically presented in Figure 3. For each of the trains, the number of intermediate carriages (N), the length of the carriages (D), the distance between the axles of the bogies (d) and the load per axle (P) are defined.

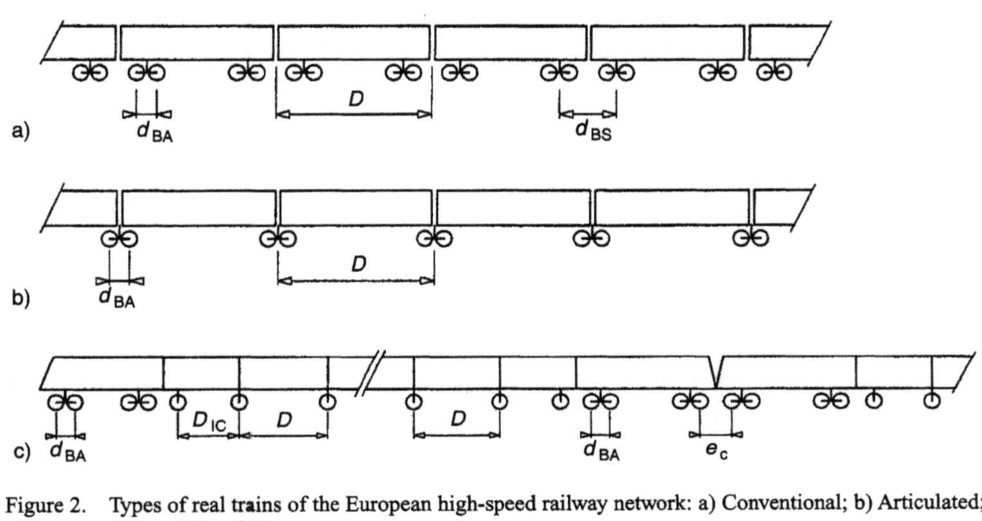

Figure 2. Types of real trains of the European high-speed railway network: a) Conventional; b) Articulated; c) Regular (adapted from [2]).

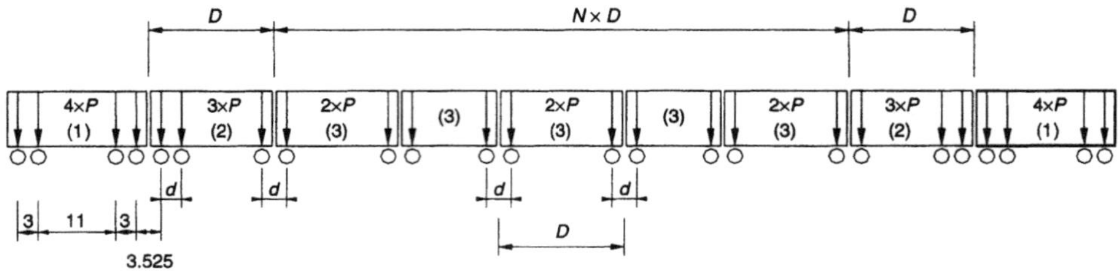

Figure 3. HSLM-A load model (adapted from [2]).

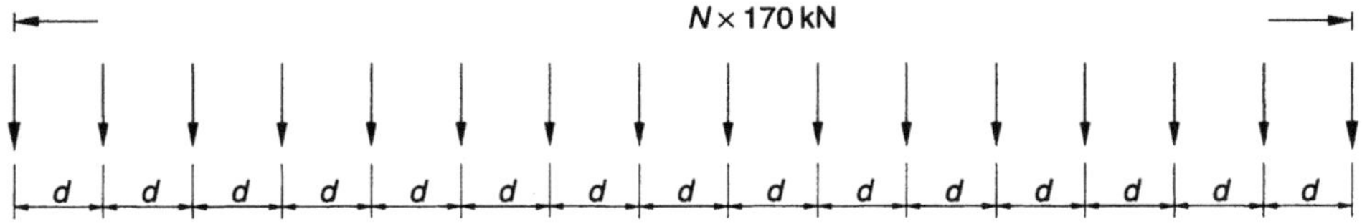

Figure 4. HSLM-B load model (adapted from [2]).

The loading model HSLM-B is composed by N point loads of 170 kN, uniformly spaced by d (Fig. 4). The values for N and d depend on the L_λ parameter, which is equal to the bridge span, L, or to half the excitation wavelength with a certain range of interest.

2.3 *Design checks*

2.3.1 *Structural safety*

Based on the results of the dynamic analyses, the dynamic amplifications should be calculated and determined by means of the following relation:

$$\varphi'_{\text{dyn}} = \max\left|\frac{y_{\text{dyn}}}{y_{\text{sta}}}\right| - 1 \tag{1}$$

where y_{dyn} is the maximum value of the dynamic response and y_{sta} is the maximum value of the static response, obtained for the passage of each of the real trains (RT) or the HSLM load model. For the bridge design, the following most unfavourable values should be taken into account

$$(1 + \varphi'_{\text{dyn}} + 0.5\varphi'') \times \begin{pmatrix} \text{HSLM} \\ \text{ou} \\ \text{RT} \end{pmatrix} \tag{2}$$

or

$$\Phi \times (LM71'' +'' SW/0) \tag{3}$$

2.3.2 *Track safety*

The deformation and structure vibration limit states, which aim at ensuring railway track safety, refer to: i) the vertical acceleration of the deck; ii) the torsion of the deck; iii) the vertical deformation of the deck and iv) the transverse deformation and vibration of the deck.

In what concerns, for instance, the vertical acceleration of the deck, excessive values of this parameter may lead to the instability of the ballast or the loss of contact between the wheel and the rail. The peak values of the vertical acceleration of the deck must not exceed 3.5 m/s^2 ($\approx$0.35 g), for the case of ballasted deck bridges, and 5.0 m/s^2 ($\approx$0.5 g), for non-ballasted deck bridges.

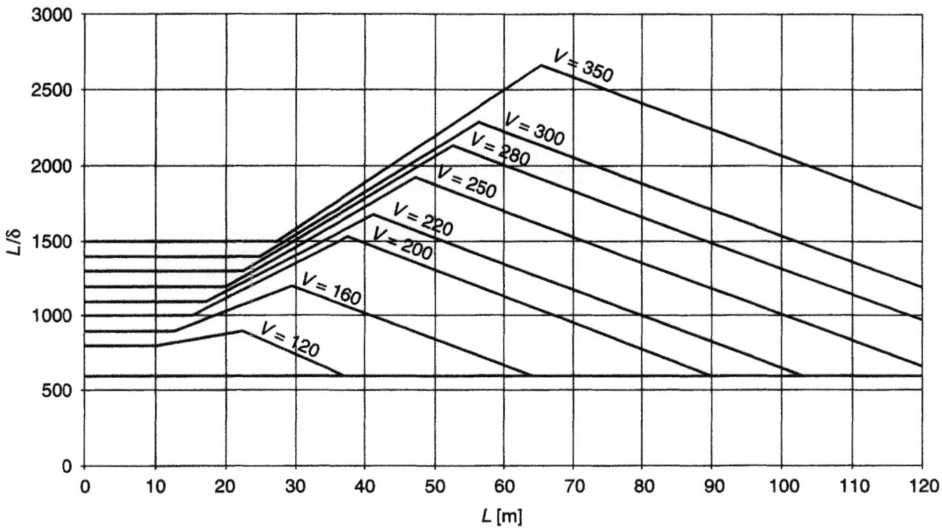

Figure 5. Maximum permissible vertical deflection δ for railway bridges with three or more successive simply supported spans corresponding to a very good level of the passengers comfort (adapted from [3]).

2.3.3 *Passengers comfort*

The comfort of the passengers depends on the accelerations to which the carriage boxes are subjected to, as the train crosses the bridge. Consequently, EN1990-AnnexA2 provides indicative values for the peak value of the vertical acceleration of the carriages (b'_v) equal to 1.0, 1.3 and 2.0 m/s^2 corresponding to three levels of comfort: very good, good and acceptable.

Hence the verification of the level of passengers comfort would involve a dynamic analysis with the bridge-train interaction, for the determination of the acceleration of the carriages and its comparison with the specified limits. For bridges consisting of simply supported spans or with continuity, which do not exhibit significant variations on the length or stiffness of its spans, and for less than 120 m spans, the verification of passengers comfort can be made in a simplified manner, by limiting the vertical displacement of the deck.

The limit values of the L/δ relation, where L is the span and δ is the maximum value of the vertical deflection of the deck, under the axle of the loaded track, for the LM71 load model majored by the respective dynamic coefficient Φ are presented, as a function of the span and the speed, for bridges in which the deck consists of three or more successive simply supported spans and for $b'_v = 1.0$ m/s^2 (Fig. 5).

For other levels of comfort, the limit values of the L/δ relation should be divided by b'_v. For decks consisting of a lower number of spans, the limit values should be multiplied by 0.7. In the case of decks of three or more successive spans with continuity, the limit values should be multiplied by 0.9.

3 DYNAMIC ANALYSIS OF A METALLIC BOWSTRING ARCH BRIDGE

The Southern Line of the Portuguese Railways (REFER) has been subject to a number of interventions towards its renovation and modernisation, and also for allowing the passage of new trains travelling at speeds higher than those to which it had been originally designed for.

With the purpose of determining the dynamic effects resulting for the passage of the high-speed "Alfa pendular" tilting train, a study of the dynamic behaviour of the bridge has been performed, which enabled to conclude a good performance of the structure up to the maximum speed reached by these trains, which is 220 km/h [6].

Figure 6. View of the 2nd span of Alcácer do Sal bridge.

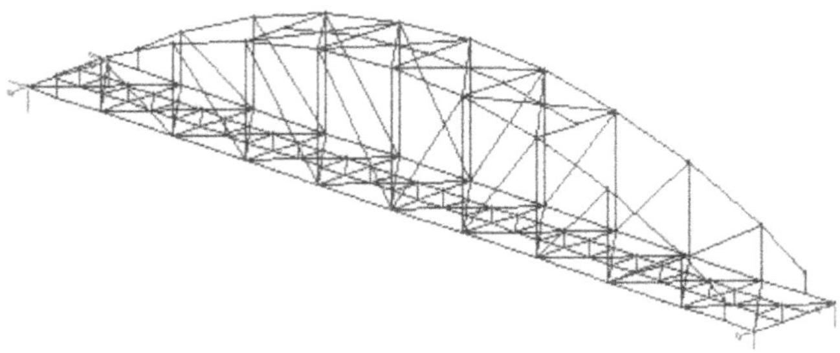

Figure 7. Finite element model of the deck.

The present work complemented the first study, by investigating the dynamic behaviour of the bridge for the passage of other trains, currently in use in the European high speed railway network [7, 8, 9].

3.1 *Description and modelling*

The Alcácer do Sal bridge is located at the km 78 + 910.240 of the Southern Line of the Portuguese Railways, where it provides the railway crossing of the Sado river. The deck, which serves as support to only one track, consists of 8 simply supported spans (14.00 + 50.04 + 49.68 + 13.41 + 3 × 50.04 + 14.00 m), in a total of 291.25 m. Each span is a steel bowstring arch, supported by masonry piers, by means of steel bearings (fixed at one end and free at the other). A side view of the 2nd span of the bridge is shown in Figure 6.

The dynamic analyses were performed based on 3D modelling, where the deck was discretized by means of 3D beam elements. The finite element modelling of one of the current spans of the deck, with 50.04 m span, is presented in Figure 7.

In Table 1 are presented the natural frequencies obtained, based on the mentioned model, as well as the physical meaning of the corresponding mode shapes.

3.2 *Dynamic analysis*

The dynamic analyses were performed for the passage of the high-speed trains VIRGIN, TALGO, TGV, EUROSTAR, THALYS, ICE2 and ETR-Y. A range of speeds from 140 km/h (≈40 m/s) to 420 km/h (1.2 × 350 km/h) was considered.

For carrying out these analyses, the method of modal superposition was used, considering the contributions of the first four vertical mode shapes and a time step Δt equal to 0.0020s. The reasons for the selection of this method were mainly related with: 1) shorter calculation time; 2) avoidance of the application of digital filters, low-pass type, to the time series of acceleration, thus reducing

Table 1. Natural frequencies and physical meaning of the corresponding mode shapes.

Mode	f (Hz)	Type of mode
1	2.58	1st transversal bending
2	5.65	1st vertical bending
3	5.83	1st torsion
4	8.20	2nd transversal bending
5	9.81	2nd torsion
6	11.69	2nd vertical bending
7	19.40	3rd vertical bending
8	26.25	4th vertical bending

the time of the results post-processing; 3) possibility of consideration of an identical damping factor for all modes of vibration.

3.3 *Results*

The maximum values of the displacement at mid-span and of the acceleration at 1/3 of the span of the deck are shown in Figure 8 as a function of the speed. The results obtained for the "Alfa Pendular" tilting train up to a speed of 265 km/h (1.2 × 220 km/h) are also presented.

3.4 *Assessment of the dynamic response*

3.4.1 *Structural safety*

Based on the dynamic analyses results, the dynamic amplifications were determined by means of the relation $\varphi'_{dyn} = \delta_{dyn}/\delta_{sta}$, in which δ_{dyn} is the peak dynamic displacement and δ_{sta} is the peak static displacement, obtained for the passage of each one of the real trains. The dynamic displacements were then corrected, to take into account the track irregularities using the relation $(1 + \varphi'_{dyn} + 0.5\varphi'') \times \delta_{sta}$. The maximum value, equal to 2.65 cm, was obtained for the TGV train at a speed of 365 km/h. This value is lower than that resulting from the application of the load model LM71 (δ_{LM71}) majored by the respective dynamic factor (Φ), equal to 3.50 cm (3.35 cm × 1.04).

3.4.2 *Track safety*

From the observation of Figure 8 can be concluded that, for the case of the "Alfa pendular" train, the limit of 5 m/s^2 ($\approx$0.5 g), is not reached of speeds up to 265 km/h. For the case of other high speed trains, it was noted that this limit was exceeded at diverse speeds, being indicated in Table 2 the lowest speeds corresponding to the occurrence of such situation (v_{lim}).

3.4.3 *Passengers comfort*

The comfort of the passengers was assessed by means of two distinct methodologies: i) based on the determination of the peak values of the acceleration of the carriages, involving dynamic analyses of the bridge-train interaction, and its comparison with the limits specified in EN1990-AnnexA2 [3]; ii) based on the simplified methodology proposed in the same standard.

The dynamic analyses of the bridge-train interaction were carried out for the Alfa pendular and ICE2 trains (Fig. 9), making use of a computer-based model, developed at the Faculty of Engineering of the University of Porto. This model considers two subsystems (bridge and train) taken as two independent structures, simultaneously calculated in time, applying an interactive process in each integration step in time domain, aiming at the compatibility of the two structural systems, in terms of the dynamic component of the interaction force and the bridge displacements under the moving loads [10, 11, 12, 13].

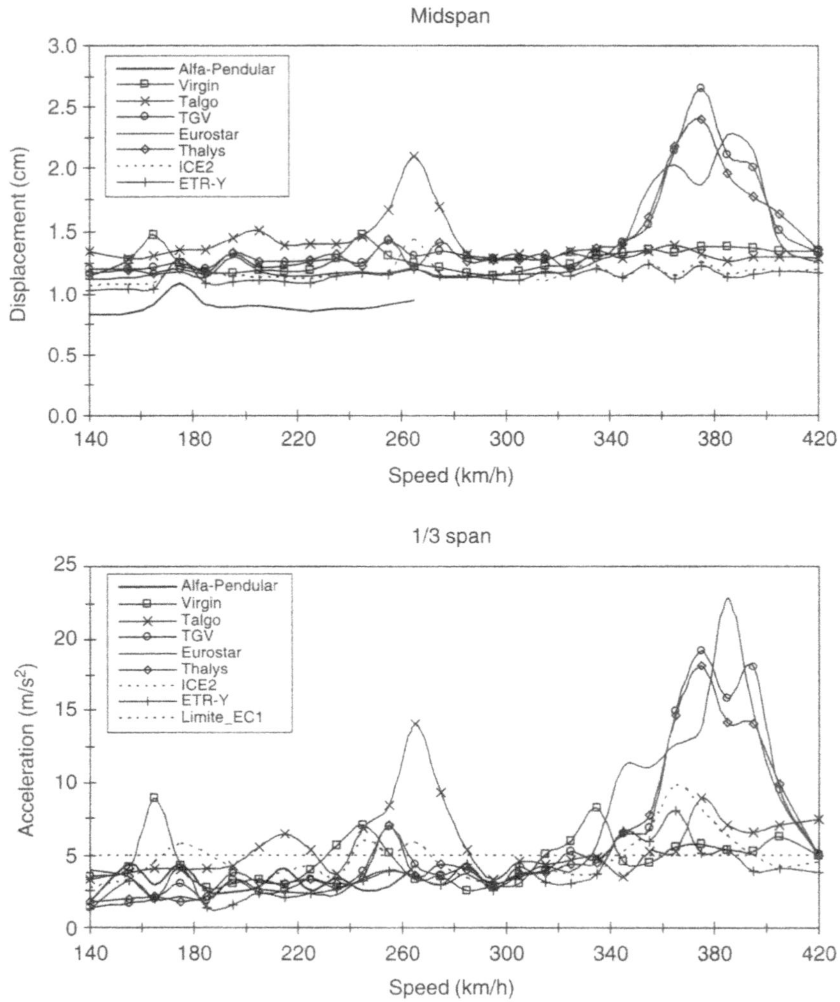

Figure 8. Maximum values of displacement at midspan and acceleration at 1/3 of the span as a function of the speed.

Table 2. Lowest speeds corresponding to the occurrence of vertical acceleration of the deck greater than 5 m/s^2.

Train	v_{lim} (km/h)
Virgin	160
Talgo	205
TGV	250
Eurostar	335
Thalys	250
ICE2	170
ETR-Y	340

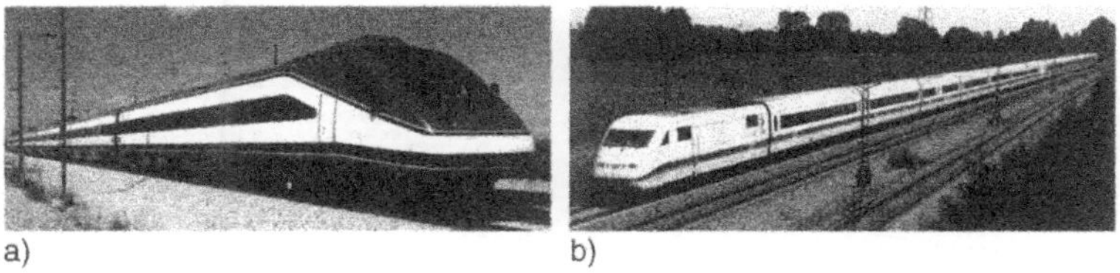

Figure 9. Trains: a) Alfa pendular; b) ICE2.

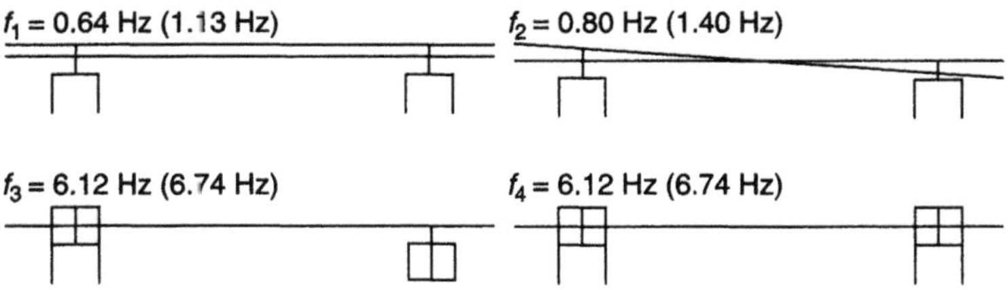

Figure 10. Representation of the mode shapes of the carriage of the ICE2 train.

These dynamic analyses with bridge-train interaction involve the consideration of appropriate vehicle models, which enable to translate its dynamic behaviour within a satisfactory approximation. From the dynamic point of view, each of the units (locomotive or carriage) can be modelled by means of the following elements: 1) a rigid body, with a mass M_c and rotation inertia I_c, to simulate the box; 2) springs, with a stiffness K_s, and dash-pots, with a damping coefficient c_s, to simulate the secondary suspensions; 3) rigid bodies, of mass M_b and rotation inertia I_b, for the simulation of the bogies; 4) springs, with a stiffness K_p, and dash-pots, with a damping coefficient c_p, to simulate the primary suspensions; 5) concentrated masses M_e, to simulate the axles and the wheels; 6) springs, with a stiffness K_h, to simulate the wheel-rail connection.

For the ICE2 train, the characteristics provided by the expert committee D214 from ERRI [4] and those referred by Paffrath [14], used on the research on the RWTH of Aachen, were adopted. As for the Alfa pendular train, the adopted model parameters were supplied by the Portuguese Railway (CP).

In Figure 10 are presented the modal configurations and frequencies associated with the first four mode shapes of the carriage of the ICE2 train, considering the parameters indicated by the D214 committee of ERRI. The first two modes, of 0.64 Hz and 0.80 Hz frequencies, involve essentially movements of the box. The other two modes, with frequencies of 6.12 Hz, correspond to vibrations of the bogies. The use of the parameters reported by Paffrath seem to lead to values of the natural frequency of the boxes, similar to those characteristics referred in the literature, which in norm are in the range [0.8;1.2] (Hz) [15].

In Figure 11 is presented the graph of the maximum values of vertical acceleration of the various carriages, as well as curves corresponding to the mean value, the 95% fractal, and the upper envelope, as a function of the speed. In the same graph are also represented the limits corresponding to the levels of comfort established in EN1990-AnnexA2. This standard sets limits to the peak value of the vertical acceleration inside the carriages equal to 1.0 m/s^2, 1.3 m/s^2 and 2.0 m/s^2 in correspondence to the three levels of passengers comfort: very good, good and acceptable.

The obtained results enable to conclude that for the ICE2 train, considering the parameters indicated by Paffrath and the upper envelope curve, the level of comfort remained very good up to a speed of about 250 km/h, good between 250 and 280 km/h, and acceptable from 280 to 340 km/h. If the 95% curve was considered, a very good level of comfort would be obtained up to a speed of about 280 km/h. For the Alfa pendular train, a very good level of comfort was obtained for the whole range of speeds analysed.

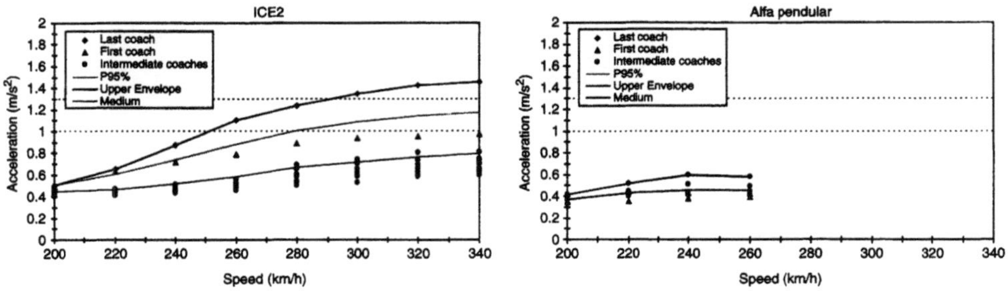

Figure 11. Vertical acceleration in the carriages of the ICE2 and Alfa pendular trains, as a function of the speed.

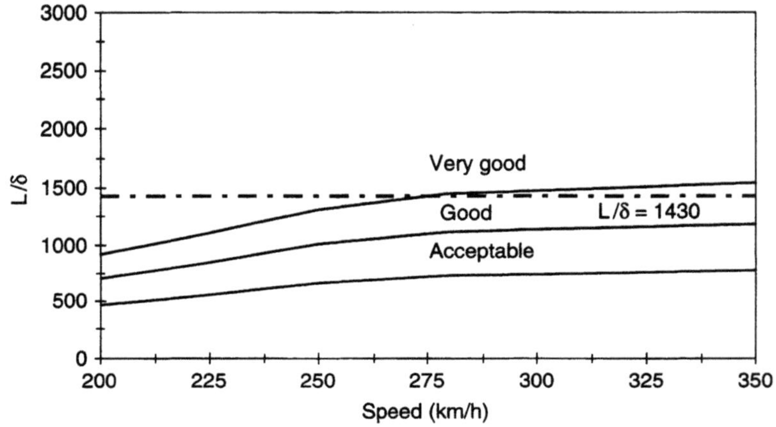

Figure 12. Identification of the levels of comfort for one single span of the Alcácer do Sal bridge.

In terms of the simplified methodology proposed in the EN1990-AnnexA2, for this bridge, $L = 50.04$ m and $\Phi \times \delta = 3.5$ cm, that is, $L/\delta = 1430$. From the observation of Figure 12 it can be concluded that the level of comfort, considering the bridge composed of one single span, would be very good for speeds up to 280 km/h, and good for higher speeds.

4 DYNAMIC ANALYSIS OF A PRESTRESSED CONCRETE VOIDED SLAB BRIDGE

One of the solutions currently used for the cases of bridge spans around 20 and 30 m is that of a deck comprising a prestressed concrete slab, voided by means of cylindrical tubes [16]. In this section is presented the study of the dynamic behaviour of a viaduct with such a deck, namely the viaduct over the Cabra river, in the Córdoba-Málaga high-speed railway line.

4.1 *Description and modelling*

The viaduct has a total length of 415.0 m, comprising a continuous deck with seventeen spans (20.0 m + 15 × 25.0 m + 20.0 m) supported by means of support elements in the piers and the intermediate columns. The deck consists of a prestressed reinforced concrete slab, which serves as support to both railway tracks (Fig. 13).

The slab is voided by means of four circular tubes of 1.2 m diameter (Fig. 14), except near the columns and piers where it is solid.

Figure 13. Viaduct over the Cabra river under construction, in the Córdoba-Málaga high-speed railway line.

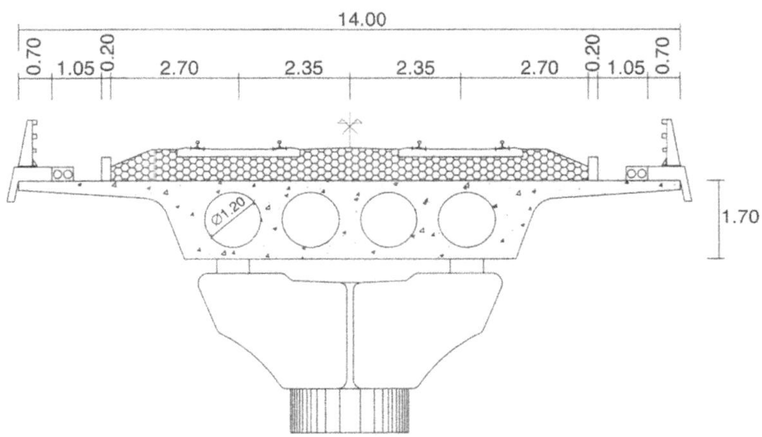

Figure 14. Cross section of the deck (voided zone).

The dynamic analyses were carried out based on a model, in which seven span of the deck were discreetized using beam elements forming a grid. In Figure 15 are presented the modal configurations of the 1st, 2nd and 3rd mode shapes, relative to the vertical bending of the deck (V), and the 8th mode, relative to the torsion of the deck (T), as well as the corresponding natural frequencies values.

For a viaduct with a prestressed reinforced concrete deck, with spans greater than 20 m, the structural damping coefficient (ζ) to be considered in the dynamic analyses should be, according to EN1991-2, equal to 1%. For spans of less than 30 m, the bridge-train interaction effects tend to reduce the peak response in resonance. Hence, for analyses where this interaction was not considered, this favourable effect has been taken into account through the introduction of an additional damping ($\Delta\zeta$). For continuous decks, the span to consider should lead to the lowest additional damping, thus corresponding to $\Delta\zeta = 0.06\%$ for $L = 25.0$ m, that is $\zeta_{total}(=\zeta + \Delta\zeta)$ equal to 1.06%.

A detailed description of this model and of its calibration based on a volumetric finite elements model, involving several comparisons between the responses obtained for the two models for specific static loads and modal parameters can be found in the reference [17].

4.2 *Dynamic analysis*

The decision on the need to perform dynamic analysis of a structure for the assessment of the dynamic effects is presented in EN1991-2 in the form of a flowchart. The decisions that led to the need to carry out a dynamic analysis of this viaduct were the following: 1) the maximum speed at

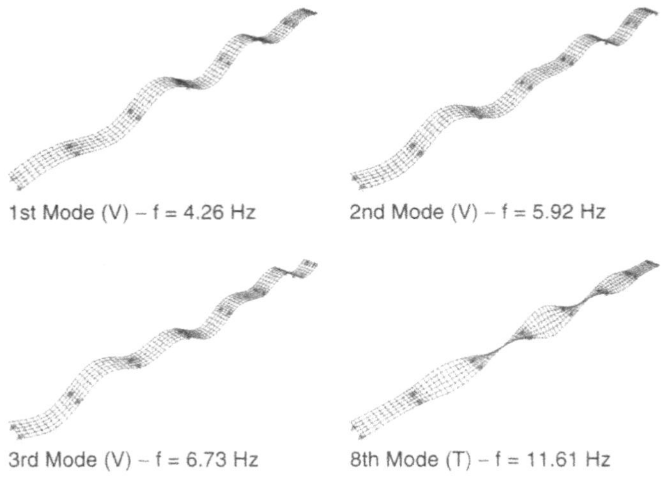

Figure 15. Modal configurations for the 1st, 2nd, 3rd and 8th mode shapes.

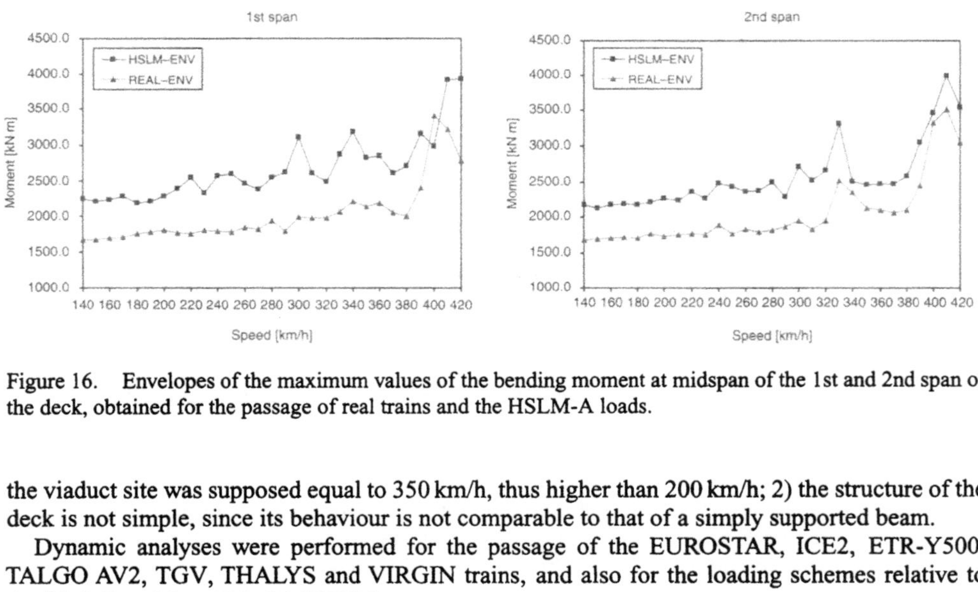

Figure 16. Envelopes of the maximum values of the bending moment at midspan of the 1st and 2nd span of the deck, obtained for the passage of real trains and the HSLM-A loads.

the viaduct site was supposed equal to 350 km/h, thus higher than 200 km/h; 2) the structure of the deck is not simple, since its behaviour is not comparable to that of a simply supported beam.

Dynamic analyses were performed for the passage of the EUROSTAR, ICE2, ETR-Y500, TALGO AV2, TGV, THALYS and VIRGIN trains, and also for the loading schemes relative to the High Speed Load Model (HSLM).

The trains were simulated as a set of moving loads, circulating along the axle of one of the tracks. For the numerical integration of the system of differential equations of dynamic equilibrium, the modal superposition method was used. According to EN1990-AnnexA2 [3], in the analyses should consider frequencies up to a maximum of: i) 30 Hz; ii) 1.5 times the frequency of the first mode shape of the structural element in analysis, including at least the first three mode shapes. For the viaduct in study, the maximum frequency is then equal to 30 Hz.

4.3 *Results*

In Figure 16 are presented the envelopes of the maximum values of the bending moment at midspan of the 1st and 2nd span of the deck, obtained for the passage of real trains and the HSLM-A

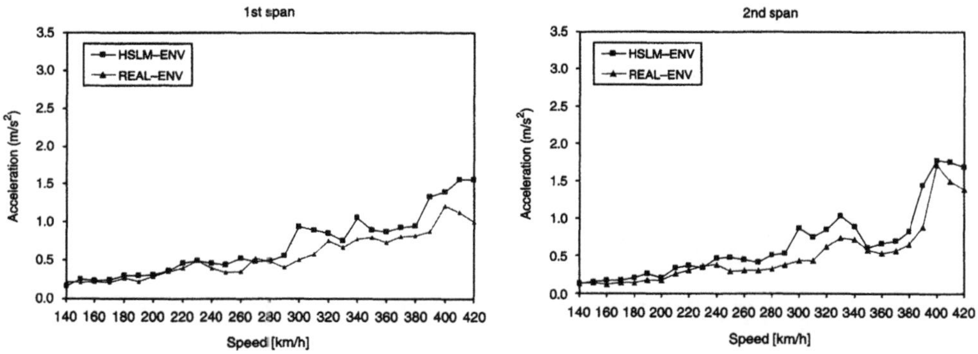

Figure 17. Envelopes regarding the maximum values of vertical acceleration at midspan of the 1st and 2nd span of the deck, obtained for the passage of real trains and the HSLM-A load model.

load model. The envelopes regarding the maximum values of vertical acceleration in the same sections, under the axle of the loaded track are in turn illustrated in Figure 17.

The observation of these graphs enables to deduce that the values obtained for the HSLM-A model are greater than those obtained for real trains, except for the bending moment at midspan of the 1st span for a speed of 400 km/h.

4.4 *Assessment of the dynamic response*

4.4.1 *Structural safety*

Table 3 compares the results regarding the maximum values of the bending moments at midspan of the 1st and 2nd span and over the 1st column, obtained by means of dynamic analysis and the application of the LM71 and SW/0 load models, majored by the respective dynamic factor. It should be noted that in the case of a deck with two railway tracks, both tracks should be loaded with the model LM71 or SW/0. For the dynamic analyses, it is only necessary to consider the loading of one of the tracks.

From the observation of the table it is possible to conclude that the bending moments obtained using the dynamic analysis are considerably lower than those resulting from the application of the LM71 and SW/0 load models. The relation between these values is of 2.8 for the 1st span, 2.9 for the 2nd span and 4.3 over the support element.

4.4.2 *Track safety*

From the graphs in Figure 18, it can be observed that the peak values of the acceleration are lower than the limit of 3.5 m/s^2 ($\approx$0.35 g), for all the cases considered.

4.4.3 *Passengers comfort*

According to EN1990-AnnexA2, for speeds greater than 350 km/h and for a very good level of comfort, the limit value for the relation L/δ, for spans between 20 m and 25 m, is equal to 1500. For continuous decks, this value should be multiplied by 0.9, that is, $(L/\delta)_{lim} = 1350$.

Looking at Table 4, it can be concluded that the displacements obtained by the application of the load model LM71, multiplied by the respective dynamic factor, are much lower than the imposed limit values.

5 DYNAMIC ANALYSIS OF A PRESTRESSED CONCRETE BOX GIRDER BRIDGE

One of the most frequently used solutions for bridges with spans around 40 m to 60 m, with prestressed concrete deck, consists on the use of a box type of cross section. This solution, not

Table 3. Maximum values of the bending moments at midspan of the 1st and 2nd span and at the 1st column, obtained by means of dynamic analysis and the application of the LM71 e SW/0 load models, majored by the respective dynamic factor (Φ).

	1st SPAN			1st SUPPORT			2nd SPAN		
	M_{sta} [kNm]	M_{dyn} [kNm]	$(1+\varphi'_{dyn}+\varphi''/2)\times M_{sta}$ [kNm]	M_{sta} [kNm]	M_{dyn} [kNm]	$(1+\varphi'_{dyn}+\varphi''/2)\times M_{sta}$ [kNm]	M_{sta} [kNm]	M_{dyn} [kNm]	$(1+\varphi'_{dyn}+\varphi''/2)\times M_{sta}$ [kNm]
Eurostar	1630.6	2758.6	2776.3	−1882.9	−2418.1	−2438.5	1575.6	3008.3	3025.4
ICE2	1533.8	1936.7	1953.3	−1737.1	−1983.1	−2002.0	1683.7	2123.4	2141.7
ETRY 500	1469.6	1848.2	1864.1	−1734.5	−1981.7	−2000.6	1598.3	2083.2	2100.5
Talgo AV2	1418.7	1950.5	1965.9	−1944.6	−2241.3	−2262.4	1478.1	1891.4	1907.5
TGV	1606.0	2966.7	2984.2	−2052.8	−2572.5	−2594.8	1546.4	3507.0	3523.8
Thalys	1600.6	3393.2	3410.6	−2051.4	−2571.5	−2593.8	1546.4	3112.6	3129.4
Virgin	1483.8	1962.1	1978.2	−2116.1	−2776.7	−2799.6	1319.2	2057.2	2071.5
HSLM-A1	1652.8	3156.8	3174.7	−1975.5	−2504.6	−2526.1	1591.7	3455.3	3472.6
HSLM-A2	1830.7	3930.5	**3950.3**	−2223.6	−2751.4	−2775.6	1767.4	3989.4	**4008.6**
HSLM-A3	1758.1	2622.2	2641.3	−2052.2	−2717.1	−2739.4	1679.2	2510.5	2528.8
HSLM-A4	1789.6	2278.0	2297.4	−2109.6	−2782.5	−2805.4	1717.2	3085.1	3103.8
HSLM-A5	1675.8	2947.6	2965.8	−1904.6	−2718.6	−2739.2	1607.0	3024.5	3041.9
HSLM-A6	1785.3	2970.2	2989.6	−2004.3	−3363.1	**−3384.9**	1720.9	2589.0	2607.7
HSLM-A7	1897.3	2614.9	2635.5	−2102.0	−3253.2	−3276.0	1841.5	2478.2	2498.2
HSLM-A8	1872.7	2719.4	2739.7	−2084.5	−2872.2	−2894.8	1835.1	2341.0	2360.9
HSLM-A9	2128.6	3177.5	3200.7	−2318.5	−3099.0	−3124.2	2100.4	2579.5	2602.3
HSLM-A10	2146.5	3102.0	3125.4	−2319.3	−3198.4	−3223.5	2138.0	2721.4	2744.7
	M_{sta}	Φ	$\Phi\times(LM71;SW/0)$	M_{sta}	Φ	$\Phi\times(LM71;SW/0)$	M_{sta}	Φ	$\Phi\times(LM71;SW/0)$
LM71 (1 track)	5122.2	1.071	5483.8	−6037.5	1.071	−6463.8	5438.1	1.071	5822.1
LM71 (2 tracks)	10244.4	1.071	**10967.6**	−12075.1	1.071	−12927.6	10876.2	1.071	**11644.1**
SW/0 (1 track)	4851.6	1.071	5194.1	−6776.9	1.071	−7255.4	5090.1	1.071	5449.5
SW/0 (2 tracks)	9703.2	1.071	10388.3	−13553.9	1.071	**−14510.8**	10180.2	1.071	10898.9

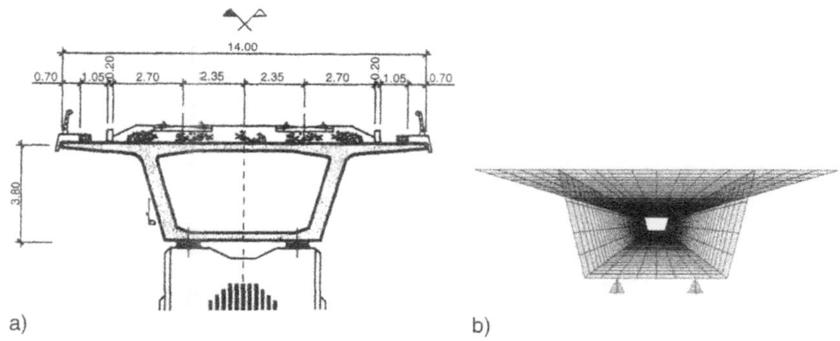

Figure 18. Cross section and finite element mesh, used for discreetizing the deck.

Table 4. Comparison of the vertical displacement of the deck for the load model LM71, multiplied by the respective dynamic coefficient Φ with the imposed limits.

Span	L = 20.0 m	L = 25.0 m
$\Phi \times \delta$LM71 (1 track) [mm]	3.8	5.7
δ lim [mm]	14.8	18.5

only leads to an excellent employment of material, but also exhibits high strength and stiffness to bending and torsion, important characteristics namely for bridges which serve as support to two railway tracks [18, 19].

The study of the dynamic behaviour of this type of bridge, designed for the Córdoba-Málaga high-speed railway line, comprising simply supported spans of 46 m, is presented in this section.

5.1 *Description and modelling*

The bridge in analysis consists of simply supported spans of 46 m, a deck on prestressed concrete, with a box type of cross section, which geometry is schematically presented in Figure 18a). The cross section height is equal to 3.8 m, corresponding to a height-span ration of 1/12. The mass of the deck per unit length (m) is equal to 40.5 t/m and corresponds to the sum of the mass of the concrete section (25/9.81 × 10.50 = 26.8 t/m) and the mass of the non-structural elements, such as the track, noise and safety barriers, ballast-guards, signals and footways, etc. (13.7 t/m).

The dynamic analyses were carried out, based on a bridge modelling in which the deck was discreetized by means of a shell finite element mesh (Fig. 18b)). In Figure 19, the modal configurations relative to the first four mode shapes of the deck can be observed, which were obtained using the above mentioned model. The corresponding natural frequencies values are also presented.

In terms of damping and according to EN1991-2, for a prestressed concrete deck bridge and a span (L) greater than 20 m, the structural damping factor (ζ) to consider in the dynamic analyses is equal to 1%.

5.2 *Dynamic analysis*

The dynamic analyses were performed for the passage of the ICE2, EUROSTAR, TGV, TALGO, THALYS, ETR-Y and VIRGIN trains, and also for the loading schemes relative to the High Speed Load Model A (HSLM-A). The trains were simulated as a set of moving loads, circulating along the axle of one of the tracks. The maximum speed at the bridge site was admitted equal to 350 km/h,

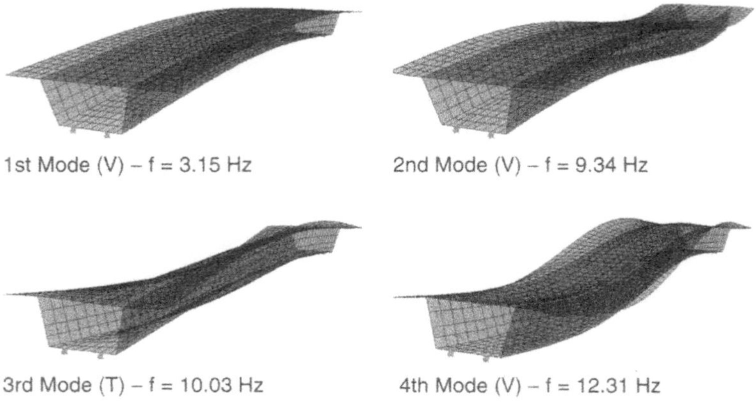

Figure 19. Modal configurations relative to the first four mode shapes of the deck.

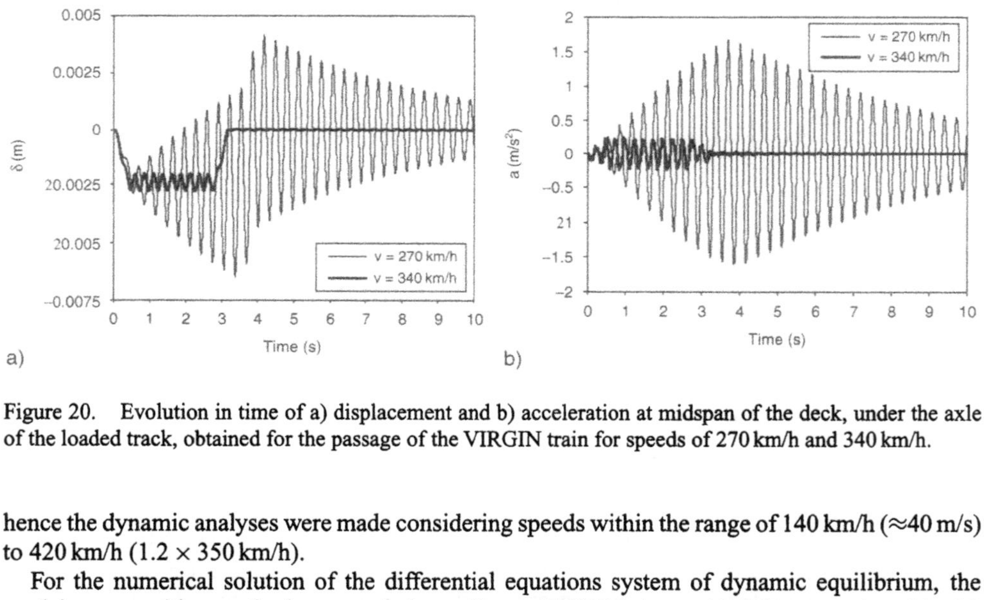

Figure 20. Evolution in time of a) displacement and b) acceleration at midspan of the deck, under the axle of the loaded track, obtained for the passage of the VIRGIN train for speeds of 270 km/h and 340 km/h.

hence the dynamic analyses were made considering speeds within the range of 140 km/h (≈40 m/s) to 420 km/h (1.2 × 350 km/h).

For the numerical solution of the differential equations system of dynamic equilibrium, the modal superposition method was used. According to EN1990-AnnexA2 [3], as mentioned in 4.2 section, the maximum frequency to be considered should be equal to 30 Hz, for the bridge in analysis.

5.3 *Results*

In Figure 20 are presented the records relative to the evolution with time of the displacement and the acceleration at midspan of the deck, under the axle of the loaded track, obtained for the passage of the VIRGIN train for speeds of 270 km/h and 340 km/h.

The maximum values of displacement and acceleration are presented in Figure 21 for the different real trains considered in the dynamic analysis, as a function of the circulating speed.

The observation of the graphs enables to conclude that the highest effects were obtained for the passage of the VIRGIN train at a speed of 270 km/h. The regular spacing between groups of axles (d) for this train equals 23.9 m. It can be noted that the resonance speed (v_{res}) detected corresponds

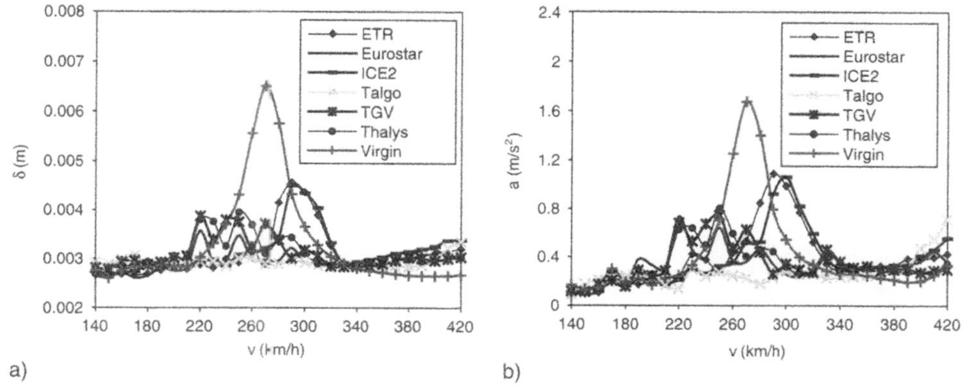

Figure 21. Maximum values of a) displacement and b) acceleration at midspan of the deck, under the axle of the loaded track, for different real trains, as a function of the speed.

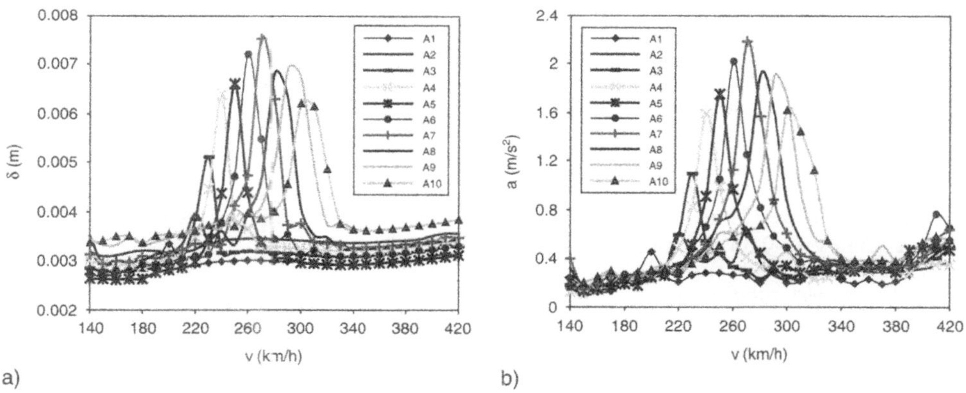

Figure 22. Maximum values of a) displacement and b) acceleration at midspan of the deck, under the axle of the loaded track, for the HSLM-A model, as a function of speed.

to the excitation of the structure with a frequency corresponding to the first mode of vibration of the bridge, that is, $v_{res} = n_0 \times d = 3.15\,\text{Hz} \times 23.9\,\text{m} = 75.3\,\text{m/s}$ ($\approx$270 km/h).

The maximum values of displacement and acceleration obtained for the passage of the different HSLM-A load schemes are given in Figure 22. In this case, the highest effects were registered for HSLM-A7 at a speed of 272 km/h. For this load scheme, the regular spacing of groups of axles is equal to 24 m. Similarly to the VIRGIN train, it is clear that this speed corresponds to the excitation of the structure with a frequency corresponding to the first mode of vibration ($3.15\,\text{Hz} \times 24\,\text{m} = 75.6\,\text{m/s} \approx 272\,\text{km/h}$).

Figure 23 illustrates the contributions of the different modes of vibration of the bridge to the responses obtained for the passage of the VIRGIN train. The curves designated by RT correspond to the total response. The other curves correspond to the isolated contributions of the first six modes of vibration to the total response.

Looking at the graphs it is possible to deduce that, in general, the response is dominated by the contribution of the first mode of vibration. For a speed of 270 km/h, to which resonance occurs in the deck, the contribution of the first mode represents about 97% of the total response, both in terms of displacements and accelerations.

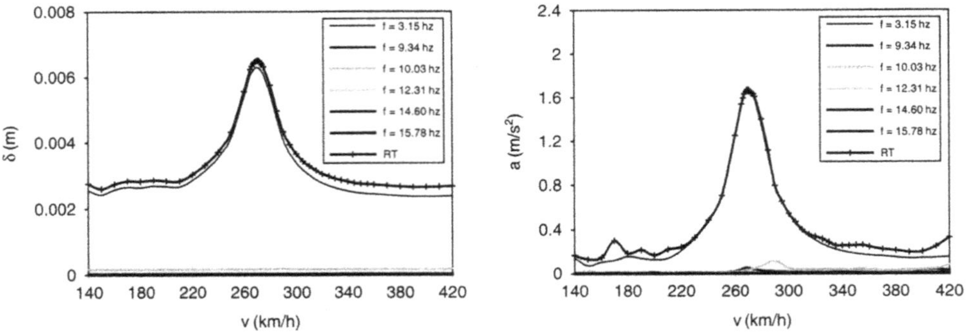

a) b)

Figure 23. Contributions of the mode of vibration to the total response in terms of maximum values of: a) displacement and b) acceleration at midspan of the deck, under the axle of the loaded track, for the passage of the VIRGIN train.

5.4 *Assessment of the dynamic response*

5.4.1 *Structural safety*

The maximum value of displacement obtained by means of the dynamic analysis was 7.52 mm, for the HSLM-A7 train at the speed of 272 km/h. This value is lower than the displacement resulting from the LM71 load model, majored by the respective dynamic factor, equal to 8.05 mm (1.04×7.75 mm).

5.4.2 *Track safety*

The graphs in Figures 21 and 22 show that the peak values of acceleration are lower than the limit value of 3.5 m/s^2 ($\approx$0.35 g) for all analysed cases.

5.4.3 *Passengers comfort*

For v = 350 km/h and L = 46 m, the lower limit indicated in EN1990-AnnexA2 for the L/δ ratio is equal to 2100. For the bridge in study, L = 46 m and $\delta = 1.04 \times 7.75$ mm = 8.05 mm, that is, L/δ = 5714, greater than 2100, which corresponds to a very good level of comfort for the passengers.

6 DYNAMIC ANALYSIS OF A COMPOSITE TWIN GIRDER BRIDGE

Along with prestressed concrete decks there are also composite solutions that have proven to be very efficient in high speed railways. One of these solutions is the composite twin girder bridge. This type of deck is used in continuous schemes with span lengths ranging from 40 m up to 65 m, and is currently a very competitive solution in France.

The use of simple structures, both in terms of structural behaviour and of easiness of assembly, associated with the use of high strength thick steel plates have made the composite twin girders very competitive in the medium spans range [20].

Composite decks are susceptible to experience high levels of vibration in resonance conditions due to its low mass and damping factor (0.5%). Complete dynamic analysis checks are therefore necessary to assess the vibration levels on the deck, the fatigue damage, and the comfort of the passengers.

6.1 *Description and modelling*

The bridge in analysis consists of a 333 m long composite twin girder deck located on the French TGV Nord line. The bridge is continuous over its entire length, comprising 7 intermediate spans of 40 m and 2 end spans of 28 m and 25 m.

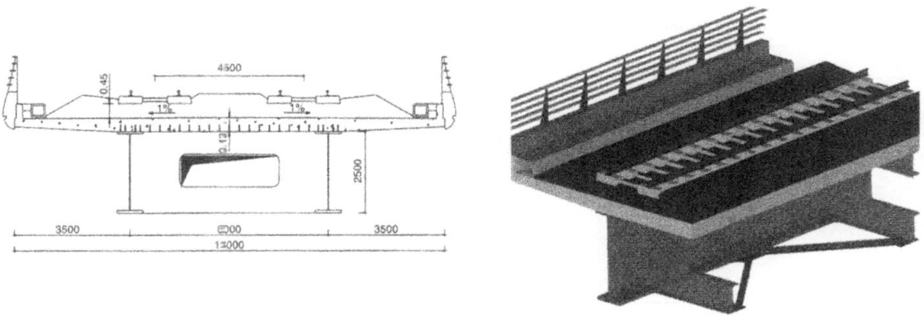

i) Front view

ii) 3D view of the deck

Figure 24. Views of the bridge deck.

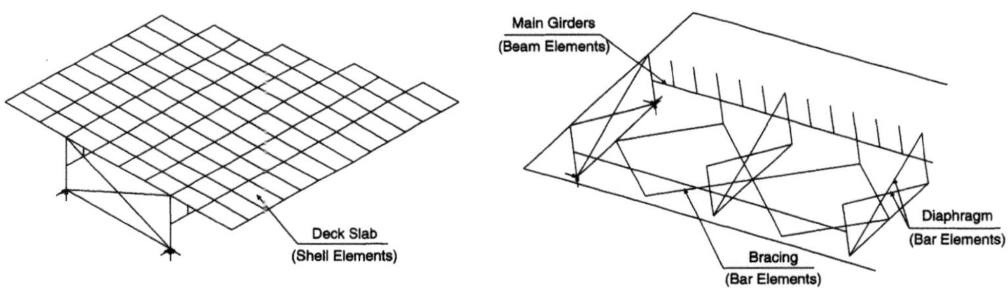

Figure 25. Numerical model of the bridge.

The main steel structure has a constant height of 2.5 m and is connected to a 0.4 m thick deck slab by stud connectors, placed both on the main and on the cross girders. These cross girders consist of plate web diaphragms and are placed every 6.667 m along the bridge length. These diaphragms reduce the deformability against distortion of the deck, induced by the eccentricity of the live load. The main girder flanges are made of longitudinally profiled plates (LP plates) over the supports, reaching a maximum thickness of 90 mm.

The deck is closed inferiorly by a horizontal steel bracing. This bracing is bolted to the bottom flanges of the main and cross girders, and provides a considerable increase of the torsional stiffness of the bridge.

Figure 24 shows the main elements of the bridge through a cross section of the deck and a 3D perspective of the right half of the deck.

Due to the eccentric position of the live loads, the torsional behaviour of the bridge should be included on the numerical model of the bridge. For this reason a 3D finite element model was chosen for the dynamic analyses.

The main girders were modelled using beam elements. Multiple sections were used in order to model the inertia variation over the supports. The deck slab was modelled using shell elements in order to reproduce shear lag effects. The composite action in the model was achieved by placing vertical stiff elements linking the main beams and the shell elements. The inferior bracing was modelled by bar elements. The diaphragm effect of the transversal elements was introduced in the model by beam elements calibrated to reproduce the in-plane deformability of the cross girders (Fig. 25).

A detailed description of the model and of its calibration based on a complete shell finite elements model can be found in reference [21].

Table 5. Comparison between calculated and identified natural frequencies.

Identified frequency (Hz)	Calculated frequency (Hz)	Type of mode
2.64	2.52	1st vertical
2.84	2.75	2nd vertical
3.12	3.09	3rd vertical
3.48	3.48	4th vertical
3.80	3.90	5th vertical
3.80	3.78	1st torsion

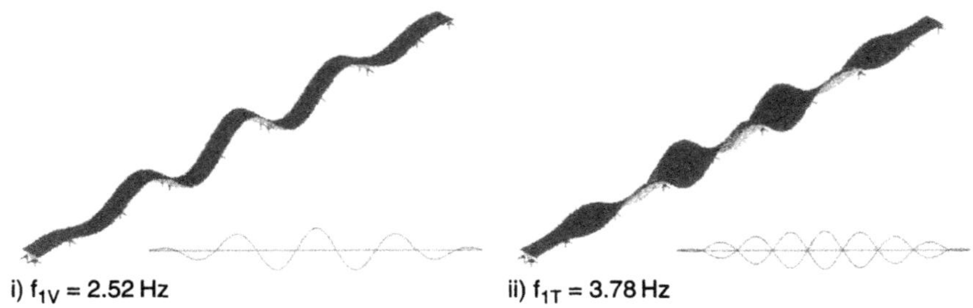

Figure 26. Mode shapes corresponding to the: i) 1st vertical bending mode; ii) 1st torsional mode.

The calculations in this paper were made considering a modular ratio of 6 ($E_c = 35$ GPa) and a non-cracked concrete section throughout the whole bridge. The mass of the deck, comprising structural elements and non-structural elements such as track and other deck accessories, was estimated to be 30 t/m.

Table 5 shows the comparison between the measured natural frequencies of the deck, referred in [22], and the correspondent values obtained from the numerical model. The mode shapes for the 1st vertical bending mode and the 1st torsion mode are shown in Figure 26.

6.2 *Dynamic analysis*

Dynamic analysis of the bridge were performed for both the European High Speed Trains (TGV, ICE 2, EUROSTAR, TALGO, THALYS, ETR 500 and VIRGIN) and the new High Speed Load Models HSLM-A (1 to 10). The analyses were conducted for speeds between 140 km/h and 420 km/h (1.2×350 km/h).

The dynamic calculations were performed using the LUSAS Bridge v13.6 software [23]. The train axle loads were assimilated to moving loads travelling across the rail lines. For the solution of the dynamic equilibrium equations the modal superposition approach was employed, each mode being analysed in the time domain using Duhamel's Integral. The analysis included modal contributions up to 30 Hz. For each mode a damping ratio of 0.5% was considered.

6.3 *Results*

Figures 27 and 28 show the envelopes of vertical displacement and vertical acceleration on the 5th and 9th spans over the loaded track, for both the real trains (RT) and the High Speed Load Models (HSLM), as a function of the speed.

From the analysis of the previous figures it can be observed that the effects induced by the HSLM are greater than those produced by the real European high speed trains, thus demonstrating the ability of the new models to reproduce the effects of high speed traffic on this type of structures.

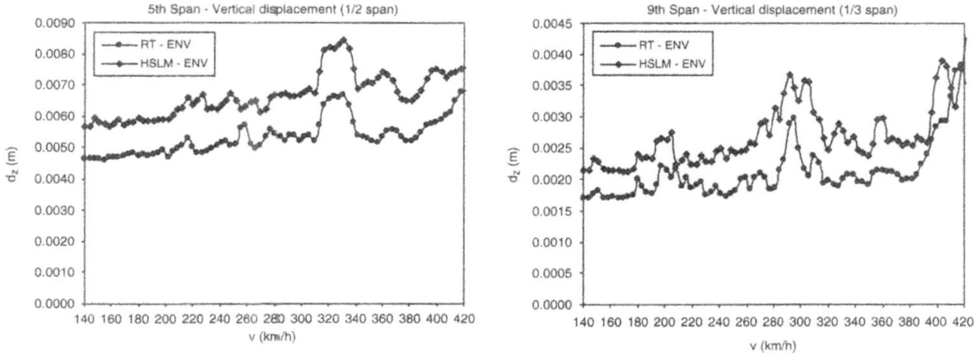

Figure 27. Maximum values of the vertical displacement in the 5th and 9th spans of the bridge, as a function of the speed.

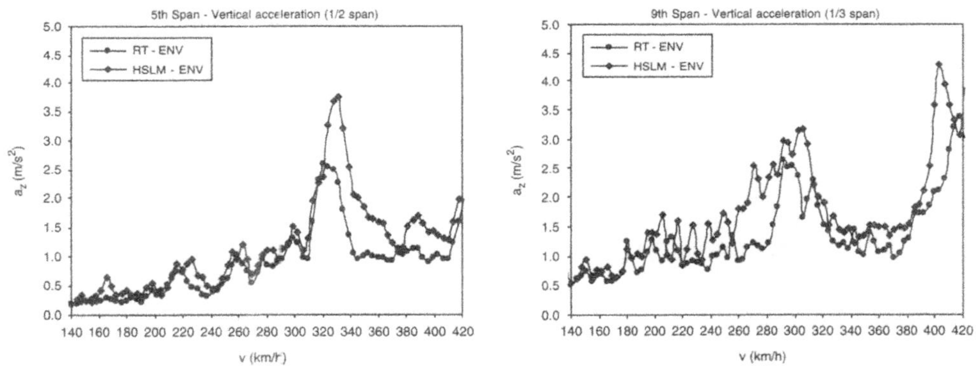

Figure 28. Maximum values of the vertical acceleration in the 5th and 9th spans of the bridge, as a function of the speed.

The highest vertical acceleration levels were registered on the two end spans, reaching a maximum value of 4.29 m/s^2 on the 9th span for a speed of 405 km/h (HSLM-A1). This peak values are associated with local modes of vertical bending and torsion. Figure 29 shows the modal contributions of a vertical bending mode ($f_V = 6.29$ Hz) and of a torsional mode ($f_T = 6.60$ Hz) for the vertical acceleration on the 9th span under HSLM-A1.

6.4 *Assessment of the dynamic response*

6.4.1 *Structural safety*

Table 6 shows the comparison between the maximum vertical deflection obtained from the dynamic analysis of the bridge and those obtained from the static calculation of the Load Model LM71, multiplied by the respective dynamic factor (Φ). In both cases, the static values of LM71 (multiplied by Φ) were greater than those obtained from the dynamic analysis.

6.4.2 *Track safety*

The vertical accelerations of the deck obtained from the analysis of the Real Trains are lower than the limit value of 0.35 g ($\approx$3.5 m/s^2) in all situations. This value is slightly higher on the 5th span with HSLM-A3 at a speed of 330 km/h (3.75 m/s^2) and on the 9th span for speeds over 400 km/h (4.29 m/s^2).

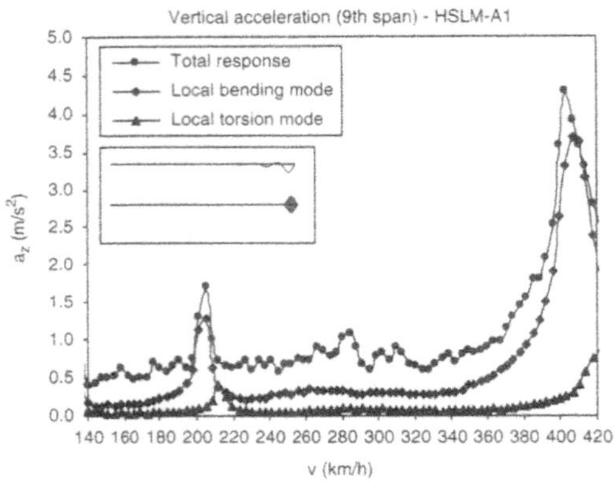

Figure 29. Contributions of local flexure and torsional modes of vibration to the vertical acceleration on the 9th span, under HSLM-A1.

Table 6. Comparison of the vertical deflection of the deck obtained from the dynamic analysis of the bridge, with the static values of the application of LM71, multiplied by the respective dynamic factor.

Span	Intermediate spans (40 m)	End spans (25 m and 28 m)
$\Phi \times$ LM71 [mm]	17.6	5.4
Dynamic analysis ("RT or HSLM") [mm]	8.5	4.3

Table 7. Comparison of the vertical deflection of the deck obtained from the application of LM 71, multiplied by the respective dynamic coefficient, with the limit values indicated in EN1990-AnnexA2.

Span	Intermediate spans (40 m)	End spans (25 m and 28 m)
$\Phi \times \delta_{LM71}$ (1 track loaded) [mm]	17.6	5.4
Limit value [mm]	18.9	15

6.4.3 *Passengers comfort*

The train vertical acceleration criterion was verified through the simplified procedure described in EN1990-AnnexA2. For a design speed of 350 km/h, the maximum permissible vertical deflections for the intermediate spans and for the end spans are, respectively, $0.9 \times L/1900$ and $0.9 \times L/1500$, in order to obtain a very good level of comfort.

The analysis of Table 7 shows that this criterion is verified for both the intermediate spans and for the end spans.

7 CONCLUSIONS

The dynamic behaviour of different types of bridges in interoperable lines, in which the Technical Specifications for Interoperability (TSI) are applicable, has been performed for both the actual high-speed real trains and the HSLM load schemes. The response of the bridges was assessed in terms of structural safety, traffic safety and passengers comfort, based on the most recent criteria established in the Eurocodes.

Four different types of bridges were analysed, the first existing in a conventional line and the others part of high-speed railway lines: a metallic bowstring arch bridge, a prestressed concrete voided slab bridge, a prestressed concrete box girder bridge and a composite twin girder bridge.

The bowstring metallic arch bridge, the Alcácer do Sal bridge, consists of 8 simply supported spans, ranging from about 14 to 50 meters in a total of 291.25 m, and only one track. This bridge belongs to a conventional line that has been subject of a number of interventions for allowing the passage of new trains travelling at speeds higher than those to which it had been originally designed for. The main goal of the study was the evaluation of the dynamic effects resulting for the passage of the high-speed "Alfa Pendular" tilting train, in order to conclude about the performance of the structure up to the maximum speed reached by these trains, which is 220 km/h, complemented with the investigating of the dynamic behaviour of the bridge for the passage of other trains, currently in use in the European high speed railway network.

The dynamic analyses were performed for the passage of the high-speed trains VIRGIN, TALGO, TGV, EUROSTAR, THALYS, ICE2 and ETR-Y, with a range of speeds from 140 km/h to 420 km/h (1.2×350 km/h), and, for the "Alfa Pendular" up to a speed of 265 km/h (1.2×220 km/h).

The maximum displacement, equal to 2.65 cm, was obtained for the TGV train at a speed of 365 km/h, significantly lower than that resulting from the application of the loading model LM71, majored by the respective dynamic factor, equal to 3.50 cm (3.35 cm $\times 1.04$).

The prestressed concrete voided slab bridge, in the Córdoba-Málaga high-speed railway, has a total length of 415.0 m, comprising a continuous deck with seventeen spans. From the structural safety point of view table it was concluded that the bending moments obtained using the dynamic analysis are considerably lower than those resulting from the application of the LM71 and SW/0 load models. The relation between these values is of 2.8 for the 1st span, 2.9 for the 2nd span and 4.3 over the support element.

The prestressed concrete box girder bridge, one of the most frequently used solutions for bridges with spans around 40 m to 60 m, included in the Córdoba-Málaga high-speed railway line, has simply supported spans of 46 m. The dynamic analysis enables to concluded that the highest effects were obtained for the passage of the VIRGIN train at a speed of 270 km/h, the resonance speed corresponding to the excitation of the structure with a frequency of the first mode shape of the bridge. The maximum values of displacement and acceleration were obtained for the passage of the different HSLM-A loading schems, with values slightly higher than the ones obtained with real trains. It was , also, concluded that, in general, the response is dominated by the contribution of the first mode shape. For a speed of 270 km/h, to which resonance occurs in the deck, the contribution of the first mode shape represents about 97% of the global response, both in terms of displacements and accelerations.

The composite twin girder deck, used in continuous schemes with span lengths ranging from 40 m up to 65 m, a very competitive solution in France, consists of a 333 m long composite twin girder deck located on the French *TGV Nord* line. The bridge is continuous over its entire length, comprising 7 intermediate spans of 40 m and 2 end spans of 28 m and 25 m.

From the analysis performed it can be concluded that the effects induced by the HSLMs are greater than those produced by the real European high speed trains, thus demonstrating the ability of the new models to reproduce the effects of high speed traffic on this type of structures. The highest vertical acceleration levels were registered on the two end spans, reaching a maximum value of $4.29\,m/s^2$ on the 9th span for a speed of 405 km/h (HSLM-A1). This peak values are associated with local modes of vertical bending and torsion.

From the track safety point of view it was concluded, in the bowstring metallic arch bridge, that the limit of $5\,m/s^2$ ($\approx 0.5\,g$) is not reached of speeds up to 265 km/h, for the case of the "Alfa Pendular" train, and, for the case of other high speed trains, it was noted that this limit was exceeded at diverse speeds. For the other three bridges the vertical accelerations of the deck obtained from the analysis of the Real Trains are lower than the limit value of $0.35\,g$ ($\approx 3.5\,m/s^2$) in all situations, except for the composite twin girder deck where this value is higher on the 5th span with HSLM-A3 at a speed of 330 km/h (3.75 m/s^2) and on the 9th span for speeds over 400 km/h (4.29 m/s^2).

For the bowstring metallic arch bridge it was observed that, in terms of the simplified methodology proposed in the eurocodes, the level of comfort, considering the bridge composed of one single span, would be very good for speeds up to 280 km/h, and good for higher speeds. When the comfort of the passengers was assessed by means a methodology based on the determination of the peak values of the acceleration involving dynamic analyses of the bridge-train interaction a slightly higher values are obtained. For the Alfa Pendular train, a very good level of comfort was obtained for the whole range of speeds analysed.

Very good level of comfort for the passengers was obtained for all the others bridges.

ACKNOWLEDGMENTS

The authors would like to thank the support of RAVE and REFER for this work and all the collaboration and information provided by Dr. Wasoodev Hoorpah, Profs. José Maria Goicolea and Felipe Gabaldón, from ETSICCyP, Universidad Politécnica of Madrid, Prof. António Reis, from GRID, Eng° José Carlos Clemente, from REFER, and Eng.° Ignacio Granell.

REFERENCES

[1] TSI (2002) – "Technical Specification for Interoperability", Official Journal of the European Communities.

[2] EN1991-2 (2003) – "Actions on Structures – Part 2: General Actions – Traffic loads on bridges", European Committee for Standardization (CEN), Brussels.

[3] EN1990-AnnexA2 (2005) – "Basis of Structural Design – Annex A2: Applications for bridges (normative)", European Committee for Standardization (CEN), Final PT Draft, Brussels.

[4] ERRI D214/RP9 (2001) – "Railway bridges for speeds >200 km/h", European Rail Research Institute (ERRI), Final Report, Utrecht.

[5] ERRI D190/RP5 (1995) – "Permissible deflections of steel and composite bridges for velocities v > 160 km/h", European Rail Research Institute (ERRI), Final Report, Utrecht.

[6] Delgado, Raimundo e and Calçada, R. (1999) – "Estudo do comportamento dinâmico da ponte de Alcácer do Sal devido a tráfego ferroviário", Relatório técnico, Instituto da Construção, Porto.

[7] Ribeiro, D., Calçada, R., Delgado and Raimundo (2005) – "Dynamic analysis of Alcácer do Sal railway bridge", Eurodyn 2005, Paris, France.

[8] Ribeiro, D. (2004) – "Comportamento Dinâmico de Pontes sob Acção de Tráfego Ferroviário a Alta Velocidade", MSc Thesis, FEUP, Porto.

[9] Ribeiro, D., Calçada, R., Delgado and Raimundo (2005) – "Interacção ponte-comboio em vias de alta velocidade: aspectos computacionais da análise dinâmica", Congreso de Métodos Numéricos en Ingeniería, Granada, Espanha.

[10] Cruz, S. (1995) – "Comportamento dinâmicos de pontes ferroviárias em vias de alta velocidade", MSc Thesis, Faculdade de Engenharia da Universidade do Porto, Porto.

[11] Calçada, R. (1996) – "Efeitos dinâmicos em pontes resultantes de tráfego ferroviário a alta velocidade", MSc Thesis, Faculdade de Engenharia da Universidade do Porto, Porto.

[12] Delgado, R. and Cruz, S. (1998) – "Modelling of railway bridge-vehicle interaction on high speed", Computers & Structures, Vol. 63, No. 3, pp. 511–523.

[13] Calçada, R. (2003) – "Avaliação experimental e numérica de efeitos dinâmicos de cargas de tráfego em pontes rodoviárias", Ph.D. Thesis, Faculdade de Engenharia da Universidade do Porto, Porto.

[14] Paffrath, P. (1998) – "Dynamisches ver halten von eisenbahnbrücken", RWTH, Aachen, Deutschland.

[15] Klaver, A. (2003) – "HSL Railway Bridges in the Netherlands", Proceedings of Conference of Structures for high-speed railway transportation, IABSE, Antwerp.

[16] Faria, I., Calçada, R. and Delgado, R. (2004) – "Análise dinâmica de um viaduto com tabuleiro contínuo em laje aligeirada em via de alta velocidade", Encontro Nacional Betão Estrutural, FEUP, Porto.

[17] Faria, I., R. Calçada, and e R. Delgado (2004) – "Comportamento Dinâmico de uma Ponte com Tabuleiro Contínuo em Laje Aligeirada sob Acção de Tráfego Ferroviário a Alta Velocidade", Congresso de Métodos Computacionais em Engenharia. LNEC.
[18] Pinto and José Rui (2005) – MSc Thesis, Faculdade de Engenharia da Universidade do Porto, Porto, Portugal (to be submitted).
[19] Pinto, J.R., Calçada, R. and Delgado, R. (2004) – "Comportamento dinâmico de uma ponte com tabuleiro em caixão sob acção de tráfego ferroviário a alta velocidade", Encontro Nacional Betão Estrutural, FEUP, Porto.
[20] Hoorpah, W., Montens, S. and Ramondenc, Ph. (2004) – "Steel and composite bridges for high speed railways – The French know-how", Bridges for High Speed Railways, Porto, Portugal.
[21] Figueiredo and Helder (2005) – MSc Thesis, Faculdade de Engenharia da Universidade do Porto, Porto, Portugal (to be submitted).
[22] Hoorpah and Wasoodev (1997) – "Contribution á l'approche numérique du comportement dynamique des ponts-rails" – Ph.D. Thesis, Université de Technologie de Compiègne, France.
[23] Finite Element Analysis Ltd (2004) – LUSAS Bridge v13.6 finite element analysis software, http://www.lusas.com.

CHAPTER 7

Dynamic behaviour of ballast on railway bridges

M. Zacher
Deutsche Bahn AG, Munich, Germany

M. Baeßler
Federal Research Institute of Material Research and Testing, Berlin, Germany

ABSTRACT: In EN 1991-2 and EN 1990 Annex A2 are set out the rules and limits for a dynamic analysis of structures. In most cases it is sufficient to model the trains as moving axle forces at constant speed. The calculated stresses have to be checked against those derived from load model Φ.LM 71. In addition the bridge deck accelerations are limited to 0.35g for bridges with ballast track. In most cases this criteria appeared to be decisive. This leads to uneconomical designs especially if the limit is slightly exceeded. Measurements on bridges showed that stiffness and damping is underestimated in many cases. A test rig at the Federal Institute for Material Research and Testing (BAM) in Berlin was built to investigate the ballast stability at high accelerations. In addition a mechanical model of the test rig was built by DB Systemtechnik in order to simulate the ballast behaviour. From both laboratory tests and simulations a modified assessment criteria for bridge deck accelerations has been derived. The modified criteria ensure safety against ballast instability and will lead to more economical designs.

1 INTRODUCTION

Vibrations of structures were excited by trains when running over bridges. In a static analysis, the increase of deformations and stresses due to dynamical effects is covered by the dynamic amplification factor Φ for load model UIC 71 and ϕ for real trains. The magnitude of the dynamic amplification factor Φ is dependent on span length and track quality. Parametric studies were carried out by expert committees ORE D23 and ORE D128 in order to determine the dynamic amplification factor. In these studies only simple supported beams and simple vehicle models were taken into account because of the lack of CPU performance and workspace capacity. For example, the structures were modelled as Euler-Bernoulli-beams and the vehicles were described by moving axle loads. Measurements on bridges have shown that the dynamic amplification factor is valid up to a speed of 200 km/h.

With the introduction of high speed trains in France on the line Paris – Lyon resonance of structures were observed, especially on those with short span length. The excessive vibrations led to a rapid deterioration of track quality and some damage of the structures. Measurements carried out by SNCF showed that bridge deck accelerations were about 1g. Obviously, in case of resonance the dynamic amplification factor might not cover the increase of deformations and stresses of structures. The interaction of train – track – bridge must be described by an additional parameter, namely the acceleration.

Investigations undertaken by SNCF showed that the ballast will loose its interlock if the bridge deck vibrations exceed 0.7g. In the framework of expert committee ERRI D214 tests with a steel box filled with ballast were carried out by the Federal Institute for Material Research and Testing [1]. The objective of the tests was to demonstrate that resilient ballast mats isolate vibrations of the bridge deck into the ballast. In order to simulate a realistic situation, 4 sleepers were embedded in 3 m long steel box filled with ballast. The sleepers were connected via UIC 60 rails. Each end of the rail was suspended elastically. The box was supported on high pressure cushions and excited via a traverse beam with a servo-hydraulic shaker. During shaking accelerations of the box and of the sleepers were recorded. The tests were carried out with accelerations up to 1g and frequencies between 2 Hz and 20 Hz. The tests showed that the amplification of the vibrations from the box to the sleeper is approximate 15%–20% without resilient mats and up to 40% with ballast mats. The test also confirmed that ballast grains will loose its interlock if accelerations exceed 0.7g.

The ERRI D214 Committee suggested rules for a dynamic assessment of bridges. These rules are now implemented in EN 1991-2 and in EN 1990 Annex 2. In order to avoid resonance effects for speeds greater than 200 km/h, bridges have to be designed dynamically[1] The acceleration limit is

- 0.35g for bridges with ballasted track
- 0.50g for bridges with ballastless track

These limits are derived by a safety factor of 2. In EN 1990 Annex 2 the frequency range is given by

$$n_{max} = \max\{30\,\text{Hz},\ 1.5 \cdot n_0, n_2\} \tag{1}$$

with n_0 is the frequency of the basic bending mode and n_2 is the third bending mode. In this context it has to be mentioned that the checks are carried out by a numerical simulation.

Note 1: Resonance is given if the exciting frequency fE multiplied with an integer k is equal a natural frequency fi of the structure. With the exciting frequency fE = v/L (L = vehicle length) the critical speeds can be calculated to the following formulae

$$v_{krit} = \frac{f_i L}{k} \quad i = 1, 2, 3, 4, \ldots \quad k = 1, 2, 3, 4 \tag{2}$$

Note 2: The safety factor for bridges with ballastless track is 2 as well. It was assumed that for accelerations above 1g contact between wheel and rail is no more guaranteed. Unfortunately this has not been proofed by numerical calculations or tests.

Measurements of vertical accelerations on axle boxes on vehicles showed that accelerations above 1g are common practice. For example, Figure 1 shows the vertical acceleration on an axle box measured on an ICE 3 train at a speed of 250 km/h on the line Hannover – Würzburg. In order to meet the frequency requirements set out in EN 1990-Annex 2 the acceleration signal is filtered low pass with a cut off frequency of 30 Hz. The maximum acceleration in this section is 2.1g.

DB Systemtechnik checked structures on the following high speed lines by a dynamical analysis:

- Erfurt – Leipzig (v_{max} = 300 km/h)
- Nürnberg – Ingolstadt (v_{max} = 300 km/h)
- Nürnberg – Ebensfeld (v_{max} = 250 km/h)
- Berlin – Rostock (v_{max} = 160 km/h)
- Köln – Rhein/Main (v_{max} = 300 km/h)
- Köln – Aachen (v_{max} = 250 km/h)
- Hamburg – Berlin (v_{max} = 230 km/h)
- Peking – Shanghai (v_{max} = 300 km/h)

The dynamic behaviour of simple supported beams, continuous beams and frames with different span length were analyzed. The simple supported beams were box girders with pre-stressed concrete, beams encased in concrete or truss girders. Most of the continuous beams were box girders with pre-stressed concrete and the majority of the frames were skewed or the track alignment was not perpendicular to the frame. Altogether more over 100 bridges were checked.

[1] In special cases a dynamic check is necessary for lower speeds

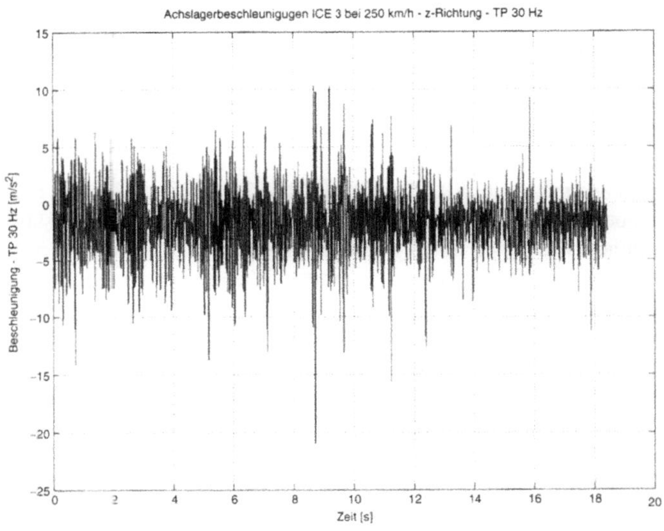

Figure 1. Vertical acceleration on an axle box of an ICE 3 with a speed of 250 km/h.

In 23 cases the acceleration criteria were not met and in 10 cases the stresses in the dynamic analysis exceeded the stresses due to load model 71. This means in 13 cases the acceleration criteria was decisive for changing the design of the structure. In most cases frames did not meet the acceleration criteria because the dynamic stiffness of the soil[2] is not available. During the passage of a train, rigid body modes were excited according to the low soil stiffness. These modes are responsible for the high accelerations.

The design of some high speed lines in Germany started long ago. At this design state a check against resonance effects was not required for structures. With the introduction of a train bridge interaction analysis in EN 1991-2 the EBA (Federal Railway Authority) demanded that the bridges shall be checked according to EN 1991-2. In some cases measurements were carried out in order to check the calculated dynamic behaviour of the bridges [5, 6, 7, 8, 9, 10]. It showed that:

- for steal beam girders encased in concrete the measured natural bending frequencies was up to 1.1 Hz–2.2 Hz higher than the predicted ones. According to Equation (2) the critical speed will be increased up to 100 km/h–200 km/h. The measured damping values were between 3.6% and 4.9%. The damping values given in EN 1991-2 are between 1.5% and 2.3% according to the span length.
- for frames the first natural bending modes were determined between 10 Hz and 25.5 Hz. Rigid body modes were not observed. It was difficult to determine damping values because the measurement signals from the strain gauges and the accelerometers were very small and free vibrations of the bridge deck could not be observed at all. Therefore damping values could only be estimated. In most cases damping was above 5%, in a special case above 10%. In EN 1991-2 no damping values are given for frames. For reinforced concrete bridges damping is between 1% and 2.2%.

With the available measurements a verification of the finite element models of the frames was possible. Now with verified mechanical models all bridges met the requirements set out in

[2] Usually in a geological expertise only the static stiffness of the soil is available. This stiffness is used for the calculation of the deformation and stresses of the frame for a predicted settlement. This means the stiffness given in the geological expertise connects the loading of the soil with its plastic deformations. In a train bridge interaction analysis dynamic stiffness values are necessary. Generally the dynamic stiffness is much higher than the static one but dynamic stiffness values are not known beforehand.

EN 1991-2 or EN 1990-A2 respectively [11]. But in a design stage this is not possible. One has to rely on the dynamic calculation. The only positive aspect is that the calculated results were on the safe side.

Again, in a dynamic train bridge interaction analysis the acceleration of the bridge deck has to be calculated for different train speeds. In order to assess the accelerations the maximum values were plotted against the train speed. An example shows Figure 2 in which the maximum acceleration of the bridge deck exceeds the limit of 0.35g at a speed of 250 km/h during the passage of an ICE 3 train. The corresponding time history plot of the bridge deck accelerations shows that the signal exceeds the limit punctually only 3 times, see marked locations in Figure 3. From a practical point of view it can be supposed that the three exceedings of the acceleration limit do not lead to a lack of interlock of the ballast stones.

In addition in a dynamic analysis frequencies up to 30 Hz have to be taken into account. As for continuous girders or for bow string bridges the first natural frequency is very low (usually 1 Hz–2 Hz). Especially for this type of bridges a lot of modes have to be taken into account although they do not contribute anything to the problem. For instance, in a dynamic analysis of a steel truss girder the first 400 natural modes were below 30 Hz. Most of the modes were local vibration of trusses or cross girders.

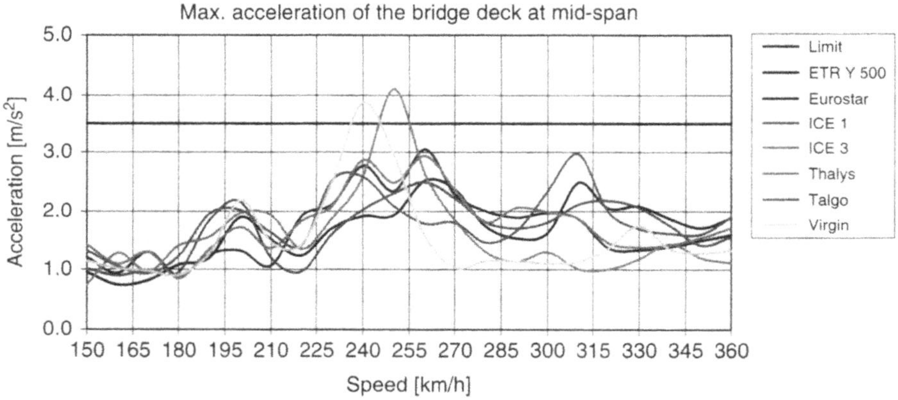

Figure 2. Maximum values of bridge deck acceleration for different trains and train speeds.

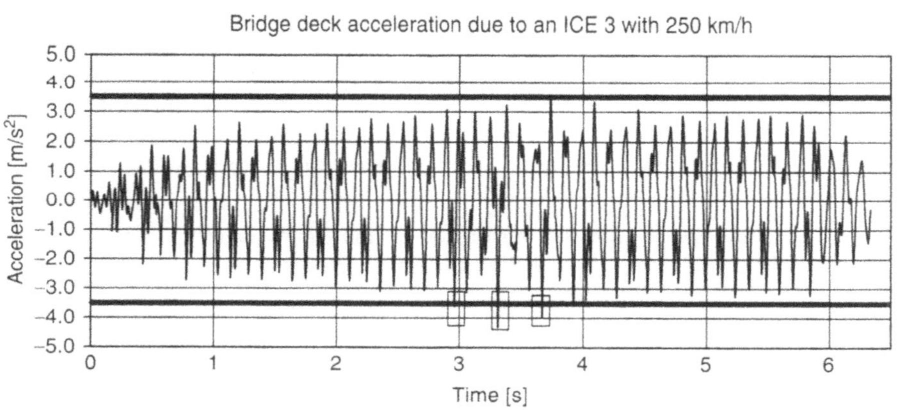

Figure 3. Time history plot of bridge deck acceleration for an ICE 3 with 250 km/h.

In order to design the bridges more economically, a modification of the acceleration criteria in EN 1991-2 and EN 1990 – Annex 2 is necessary. The question is how much and how frequently can the current acceleration limit exceeded without taking risk of loosing interlock between the ballast stones. In order to find the answer of the question, DB Systemtechnik started a co-operation with BAM.

2 LABORATORY TESTS AT BAM

Since it is not possible to investigate the ballast behaviour at higher frequencies at the big test rig designed for the ERRI D214 committee, a smaller test rig was built. Figure 4 shows a principle sketch (left) and a photo (right) of the test rig. The dimension of the base area of the steel box is 1.05 m × 1.05 m. The box is supported with a servo-hydraulic shaker and guided with rails on both sides of the box. The shaker is able to excite the box to vertical vibrations up to 2g and to frequencies up to 60 Hz. The box is filled up with ballast. A concrete bloc (0.3 m × 0.2 m × 0.4 m) representing the sleeper is embedded in the ballast. The thickness of the ballast is 35 cm. The sleeper can be loaded by a second servo-hydraulic shaker.

Accelerometers are fixed upon the sleeper and at the corners of the box. In addition settlements of the sleeper are measured with inductive transducer. Between sleeper and shaker a force measurement gauge is fixed in order to control the forces upon the sleeper. A detailed description of the test rig can be found in [2].

The result of the test is strongly dependent on the initial state of the configuration. Therefore it is important that the initial state is approximately the same for all test series. Especially the porosity of the ballast is an important parameter. Since the porosity can hardly be determined, the box was exited with vibrations of 1g and a frequency of 1 Hz as long as the settlements of the sleeper did not change significantly. After shaking the sleeper was loaded with 20 kN with a frequency of 1 Hz in order to compact the ballast.

2.1 *Transfer function of ballast*

In order to compare the results of the small test rig with those of the test rig used in ERRI D214, the transfer function between box and sleeper (sleeper acceleration/box acceleration) was determined. Therefore the box was exited harmonically and the accelerations of sleeper and box were recorded.

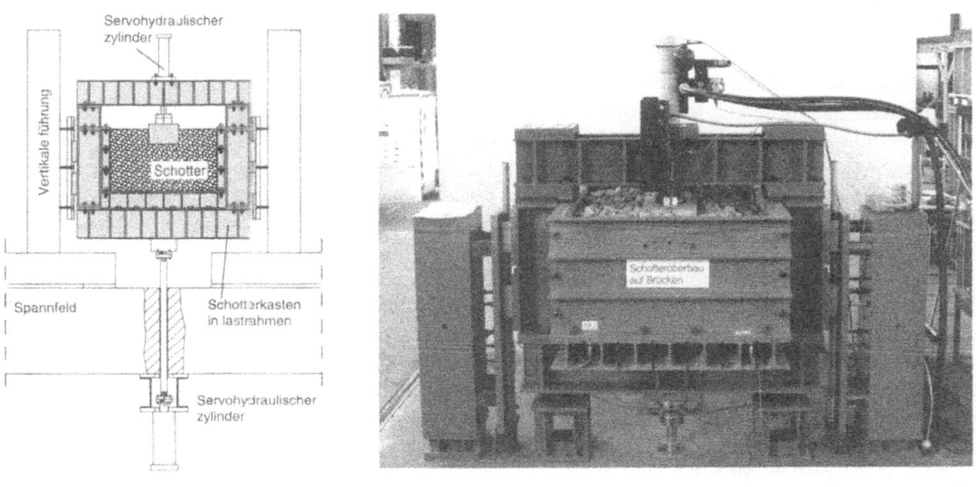

Figure 4. Test rig at the BAM [2].

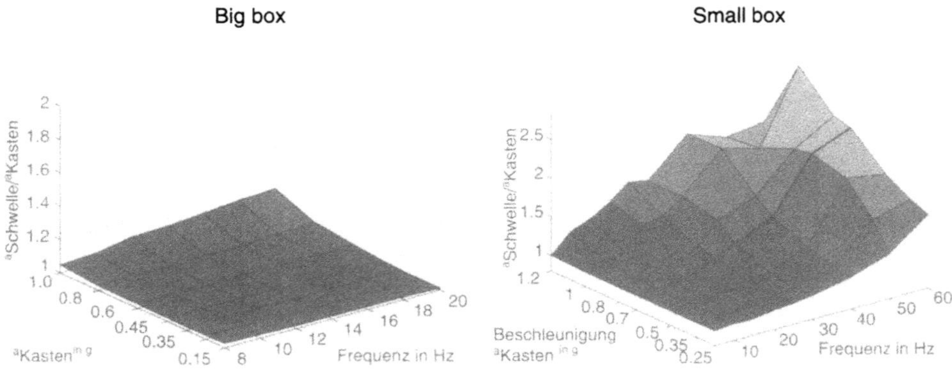

Figure 5. Transfer function sleeper/box [1, 2].

Figure 5 shows both transfer functions dependent on the amplitude and frequency of the box acceleration. The diagram on the left hand side shows the transfer function determined at the big box. Again, with this test rig amplitudes of 1 g and frequencies up to 20 Hz were considered. The diagram on the right hand side shows the transfer function determined at the small box. A comparison of both transfer functions shows that they match very well in the interested frequency range. That means the small test rig produces comparable results and can be used for further investigations.

Figure 5 shows that the transfer function increase with the frequency. The red line in the right diagram indicate the excitation amplitude of 0.7g (critical state).

2.2 *Settlement of the sleeper*

In order to estimate the settlement behaviour of the sleeper and the correlated maintenance work, that would be necessary in case of high bridge deck vibrations, additional test were carried out at the test rig. After the ballast was compacted in the same way described above, the sleeper was loaded with a frequency of 1 Hz, see Figure 6. During this phase the box did not move. The maximum load was chosen to 20 kN because this load causes the same pressure on the ballast surface as a loco with an axle load of 20 t standing in a real track.

Afterwards 50 vibration cycles of the box (acceleration a_B with a constant frequency) alternated with 20 load cycles of the sleeper. This procedure was repeated 20 times. Then the sleeper was loaded again with 10^3 cycles. The amplitude of the box acceleration was increased and the whole procedure started again. After the test with the highest acceleration was finished the ballast was compacted once more and the whole series of tests with a higher frequency was repeated. During the tests the settlement of the sleeper was measured continuously.

Figure 7 shows the sleeper settlements dependent on the number of load cycles N. The settlements are plotted for different frequencies. In order to compare the magnitude of the settlements of each test, the settlements in the diagram are normalized at a value of 9000 load cycles because the initial settlements differ in each series of tests. The repetitive shaking and loading sequences are shaded in grey. Above the grey columns the excitation amplitude is indicated. One observes that for higher accelerations the settlement of the sleeper increases considerably because the loss of interlock is proportional to the acceleration of the box. It is quite clear that the greater the loss of interlock the greater the settlements during cyclic loading.

The settlement behaviour of the sleeper is also dependent on the frequency. Larger settlements values are observer at higher frequencies. The greatest settlements are at a frequency of 40 Hz. On the other hand at a frequency of 50 Hz the settlements are even smaller. In order to get a clearer picture of the settlement behaviour depending on the frequency more tests will be necessary. This was not possible in the framework of this project.

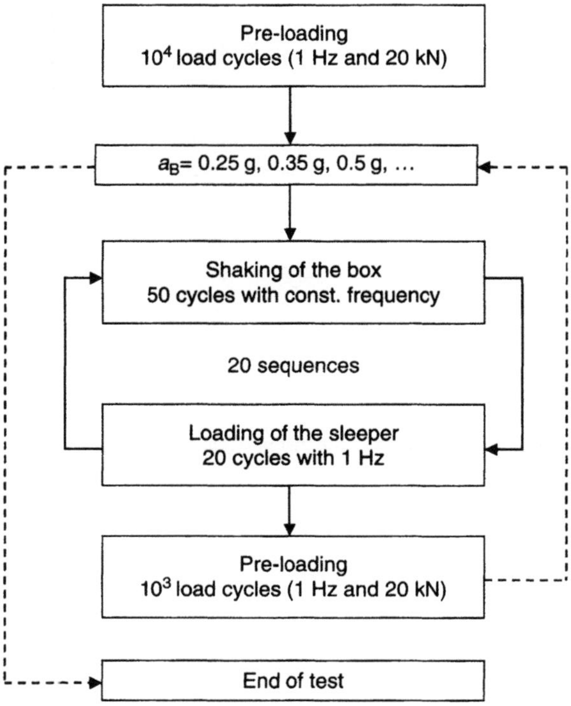

Figure 6. Test campaign for settlement behaviour [2].

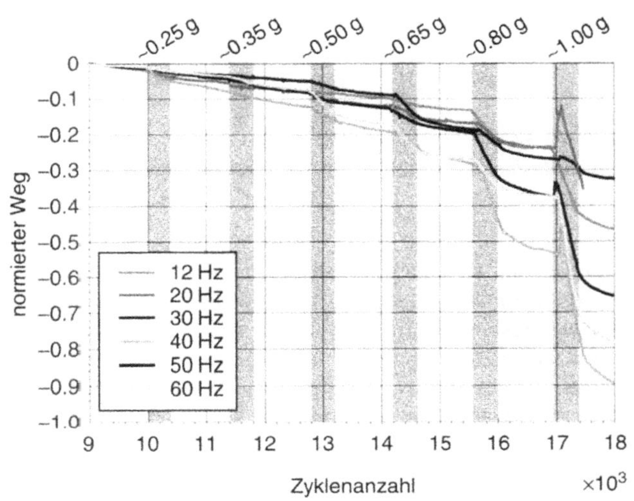

Figure 7. Settlement of the sleeper dependent on acceleration and frequency [2].

2.3 *Lateral resistance of the sleeper under vibrations*

All laboratory tests showed that the ballast grains will loose their interlock at accelerations of 0.7g. As for amplitudes below 0.7g the effect on the ballast is relatively small. In order to assess the "stability" of the ballast against accelerations, test have to be carried out in an ultimate state. In

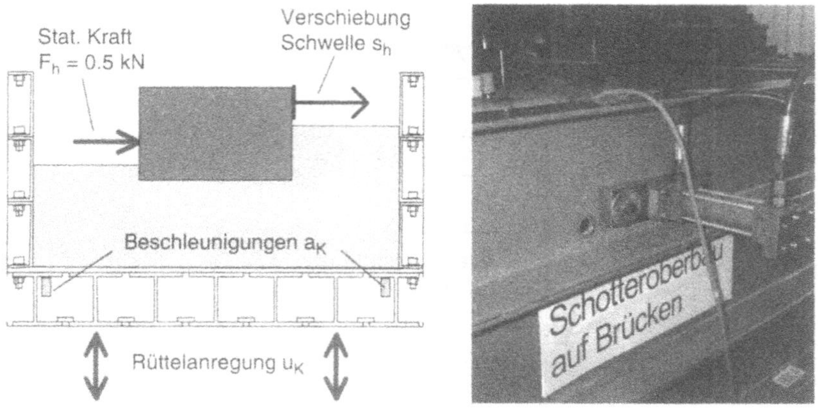

Figure 8. Principle of the test (left) and hydraulic piston (right) [2].

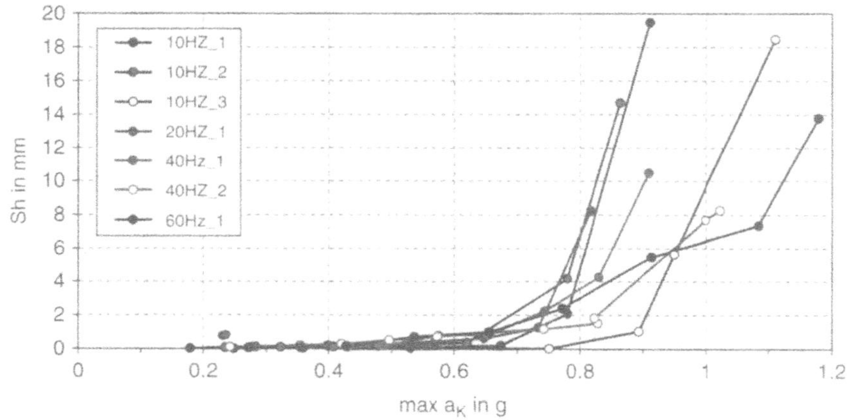

Figure 9. Lateral displacements of the sleeper after 500 acceleration cycles at different amplitudes and frequencies[3] [2].

addition it is necessary to find an objective assessment criteria for an instable ballast behaviour. One suitable indicator of the quality of the interlock between the grains is the lateral resistance of the sleeper. If the grains loose its interlock the lateral resistance of the sleeper will tend to zero. Analogous to lateral resistance measurements on sleepers in the track the sleeper in the test rig was loaded with a lateral force of 0.5 kN, see Figure 8. During the test the lateral displacement of the sleeper was measured with an inductive transducer. The magnitude of the lateral displacement was used as an assessment criteria for the loss of interlock between the ballast grains. The box was excited with 500 acceleration cycles with different amplitudes and frequencies. During one test the amplitude and the frequency was kept constant. After each series of tests the initial state was rearranged due to the method described above.

In Figure 9 the lateral displacements of the sleeper after 500 acceleration cycles are plotted against the acceleration amplitudes at different frequencies. The curves in the diagram look like an e-function. Below an acceleration amplitude of 0.7g the sleeper hardly moves. Above 0.7g the

[3] The last figure in the legend describes the test number.

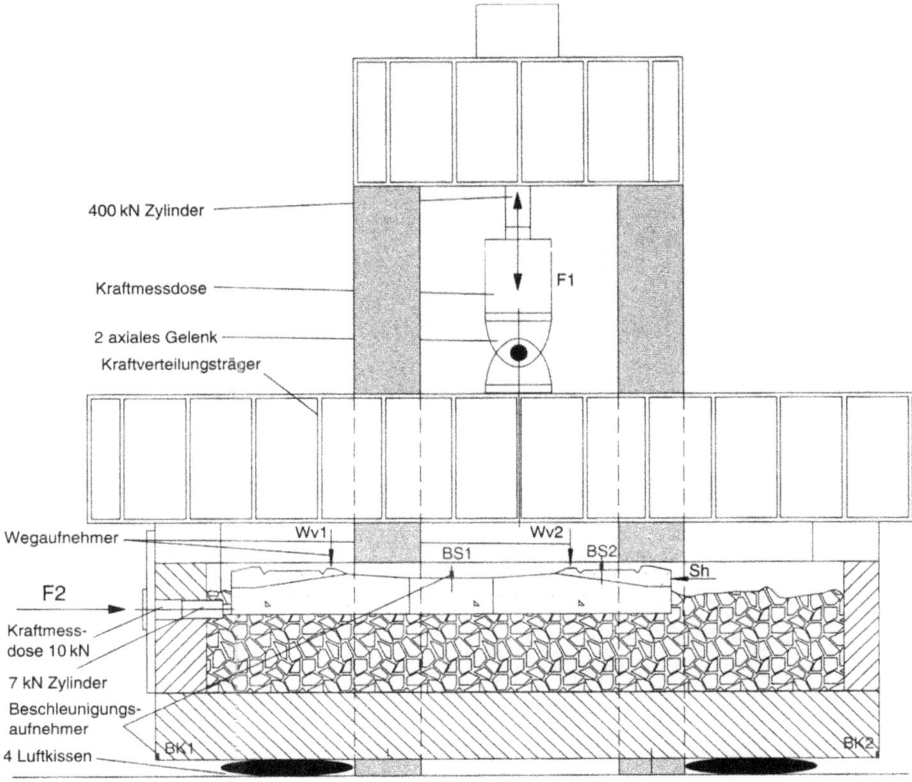

Figure 10. Sketch of the big test rig [2].

lateral displacements increase very fast. Astonishingly the lateral displacements are not dependent on the frequency.

Since the small test rig was not originally designed for lateral resistance test during shaking, problems during the tests occurred. Some tests had be stopped before completion because the concrete bloc tilted or started to turn over. In order to verify the results, the tests were repeated once more at a larger system.

Another steel box was manufactured. The steel box was 300 cm long and 92 cm wide. A sketch of the test rig shows Figure. 10. A photo of the whole site shows Figure 11. The B 70 sleeper was shortened at both ends to a total length of 2.1 m in order to ensure an 80 cm thick ballast bed between the shoulder of the sleeper and the box. Because of the length of the sleeper tilting is no longer possible. The box is supported on high air pressure cushions and is excited via a traverse beam with an servo-hydraulic shaker (Force F1).

Before the tests started the ballast had to be compacted. This was achieved through shaking the box with 1g. Afterwards the sleeper was loaded with 1000 cycles with a force between 2 and 70 kN. The series of tests were carried out in the same way as the tests at the small test rig. A lateral force of 1 kN (Force F2) was applied on the sleeper during shaking. The lateral displacement of the sleeper (Sh) was measured continuously. After 500 vibration cycles the lateral displacement was noticed. Afterwards the lateral resistance of the embedded sleeper was determined statically without shaking. According to DB rules the lateral force is assessed at a sleeper displacement of 2 mm. The measured values were between 4.5 kN and 7 kN. These values comply with values determined on B70 sleepers in a track which was tamped beforehand.

In Figure 12 the lateral displacements of the sleeper after 500 vibration cycles are plotted against the acceleration amplitudes at different frequencies. The dotted lines show the result with resilient

Figure 11. Big test rig in the laboratory hall [2].

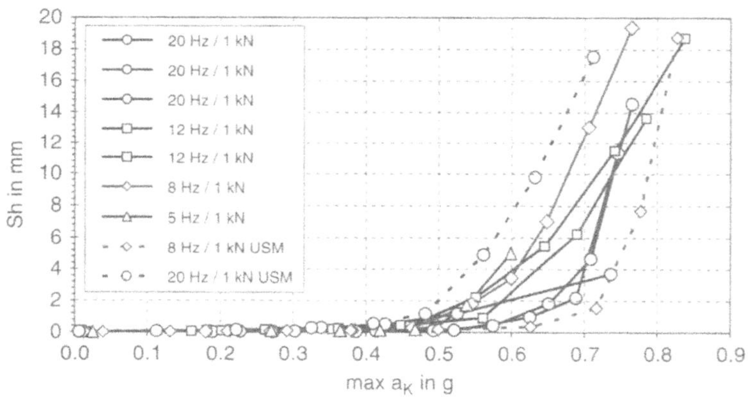

Figure 12. Lateral displacements of the sleeper after 500 acceleration cycles at different amplitudes and frequencies [2].

ballast mats. A comparison with Figure 9 shows that the maximum displacements are the same. Also at this large system the sleeper displacements are relatively independent of the frequency.

3 MODELLING

In addition a mechanical model of the ballast has been developed by DB Systemtechnik with the commercial computer program PFC^{2D} (Particle Flow Code 2D). In PFC^{2D} the ballast grains are represented as balls. It is assumed that the grains are rigid. Each particle has 3 degrees of freedom (2 translational and 1 rotational DOF). The laws of motion (Newton-Euler) are solved by numerical integration for each particle. A finite difference integration scheme is used in PFC^{2D}.

In PFC^{2D} several contact laws are implemented. The simplest contact model is a non-linear spring transmitting only compression forces. This constitutive contact law was applied for normal and tangential contact forces. The tangential force is limited according the Coulomb law ($F_T \leq \mu \cdot F_N$). Viscous dampers act in parallel to the contact springs in order to model the loss of energy. Usually the microscopic parameters (contact stiffness, contact damping, coefficient of friction) are not known. For the verification of the model the microscopic parameters had to be adjusted as long

Figure 13. Mechanical model of the test rig.

as the measured macroscopic parameters (deformations, forces) are met. Some values for contact stiffness and damping are given in [3].

In PFC^{2D} the modelling of rigid bodies with arbitrary geometry is complicated. The geometry of rigid bodies has to be approximated by small balls that are linked together by "parallel bonds". A parallel bond acts as a glue or a cemetatious material that is able to transmit both forces and moments. A parallel bond acts in parallel to the constitutive contact laws mentioned above.

The mechanical model of the box, sleeper and ballast was built by means of the deposit method. This is done by means of the following steps.

1. In a first step the ballast grains are generated randomly above the base area of the box. In this stage overlapping of grains is not permitted because -according to the high contact forces-the grains would fly away immediately at the start of a simulation. After the generation of particles a numerical simulation starts. Due to gravitation the grains fall into the box and they organize themselves dependent on the constitutive laws. The simulation ends if the sum of kinetic energy of the particles is below a small limit. Then a top ballast surface was generated by means of deleting all grains which centre of gravity is above 35 cm.
2. Now the sleeper is modelled above the ballast surface. Small particles were arranged in a way that their surfaces touches each other. At the contact points parallel bonds were defined.
3. In addition new ballast grains were generated between the sleeper and the walls of the box. A second numerical simulation is necessary in order to arrange the ballast grains between sleeper and wall.
4. Similar to the laboratory test the whole box was excited with vertical vibration of 1g and afterwards the sleeper was cyclic loaded in order to compact the ballast. In contrast to the laboratory test a fraction of load cycles can be applied in the simulation model because of the large amount of computing time.

Figure 13 shows the mechanical model of the box. The red balls represent the sleeper the blue ones the ballast. The black lines represent the forces acting between the ballast grains. The thickness of the line is proportional to the magnitude of the force. Table 1 show the data used in the simulation. The damping value is adopted from Kruse [3]. The friction coefficient between the ballast grains was estimated.

Figure 14 shows the measured and calculated transfer function. The rms-values of the relation of sleeper to box acceleration are displayed. The box was excited with 0.7g. It can be observed that the transfer function has a maximum at 60 Hz.

Table 1. Data used in the simulation

Normal contact stiffness [N/m]	$4.0 \cdot 10^8$
Shear contact stiffness [N/m]	$1.0 \cdot 10^9$
Normal stiffness of parallel bond [N/m]	$1.0 \cdot 10^{10}$
Shear stiffness of parallel bond [N/m]	$1.0 \cdot 10^{10}$
Value of Critical damping [/]	0.5
Friction coefficient [/]	0.9
Density of a ballast grain [kg/m³]	2800
Density of the sleeper [kg/m³]	2500

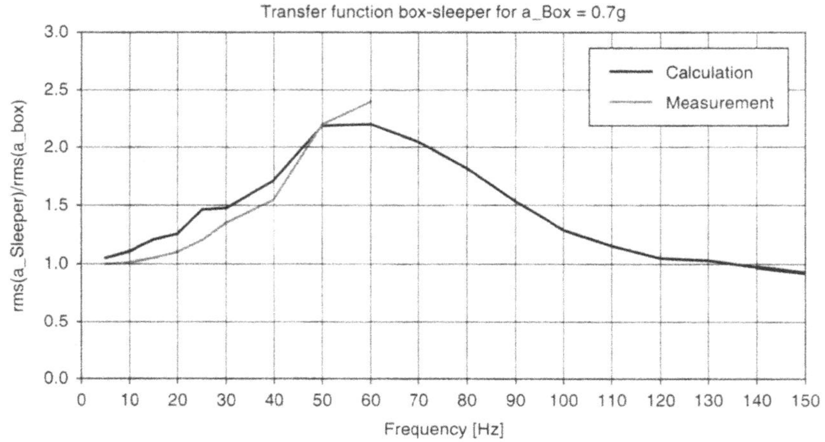

Figure 14. Measured [2] and calculated transfer function.

If one consider the fact that

- the simulation model is only 2 dimensional,
- an interlock of balls is hardly possible,
- the initial state of the ballast (porosity, stresses) in the simulation model differs from that in the laboratory test
- the results of the measurements are obtained form one series of test only and the scatter is not known
- the correspondence between measurement and simulations is quite good.

4 IMPROVEMENT OF THE ACCELERATION CRITERIA

Based on the in situ measurements, dynamic calculations and the laboratory test, DB Systemtechnik has worked out an improvement for the acceleration criteria, which is described in this paragraph. All investigations undertaken at Deutsche Bahn AG showed, that

- the calculated bridge deck accelerations are usually overestimated in a dynamic analysis in a design state
- for bridges where the bridge deck acceleration slightly exceeds the limits set out in EN 1990-Annex 2 no deterioration of the track was observed.

In addition the laboratory test carried out at the BAM showed that the structure of the ballast is not affected significantly for bridge deck accelerations below 0.7g. Due to the lateral resistance tests described above the lateral sleeper displacement below 0.7g is also very small. In most of the tests the measured lateral sleeper displacements increase linearly with the number of cycles [2].

According to the knowledge gained in the recent years it seems good engineering practice to determine an additional acceleration criteria derived from the lateral resistance test. If the bridge deck acceleration exceeds the limit set out in EN 1990-Annex 2, a certain number of cycles shall be permitted until a lateral sleeper displacement of 0.5 mm is reached. The value of 0.5 mm is a conservative limit. For example, in in situ tests for determining the lateral track resistance an increment of the track shift of 1 mm is permitted [4]. In this case a smaller limit was chosen because in laboratory tests the boundary conditions (porosity, frictions coefficient) can be controlled more easily than in real track.

From Figure 9 and Figure 12 can be derived that below 0.35g the maximal lateral displacement of the sleeper is below 0.5 mm. The test showed that for high bridge deck accelerations the maximum lateral sleeper displacement after 500 vibration cycles are approximately 20 mm. From this data a displacement increment of 0.04 mm per cycle is calculated. In this case 10 cycles of high accelerations can be permitted. In the laboratory tests big lateral sleeper displacement occurred only at accelerations above 0.7g. So it would be not good engineering practice to allow bridge deck acceleration at the stability limit of the ballast. Therefore it is proposed to reduce the above mentioned safety factor from 2 to a value of 1.3. The same safety factor is given in DIN 1054:2003-01 for the ultimate state of the loading capacity of the soil. With a safety factor of 1.3 a maximum acceleration of 0.55g can be permitted.

According to Figure 14 the maximum value of the transfer function occurs at a frequency of 60 Hz. At this frequency there is a maximal amplification of the acceleration from bridge deck to sleepers. Therefore in a dynamic analysis only those modes shall be taken into account whose natural frequencies are below 60 Hz. Generally the natural frequency of the first bending mode is far below 60 Hz. In most cases it is sufficient to take the first 3 natural modes into account that can be excited by traffic loads.

The axle box accelerations in Figure 1 shows that accelerations above 1g are not safety relevant. But the loading of ballastless structures will increase, if vibrations of 1g are permitted. Since no investigations on fatigue under high accelerations are undertaken, the above safety factor shall be applied for ballastless bridges.

The following recommendations are summarized below:

1) Usually the max. permitted bridge deck accelerations is

- 0.35g for bridges with ballast track
- 0.50g for bridges with ballastless track

2) In exceptional cases it is permitted to exceed the above values for 10 successive vibration cycles. In this case the max. acceleration is limited to

- 0.55g for bridges with ballast track
- 0.75g for bridges with ballastless track

In a dynamic analysis the frequency range shall be defined according to the formula:

$$n_{max} = \min\{n_3,\ 60\,\text{Hz}\} \tag{3}$$

with n_3 is the natural frequency of the 3rd natural mode that can be excited by traffic loads.

REFERENCES

[1] Rohrmann, Rolf 1998. Experimentelle Untersuchungen des Schotterverhaltens auf Brücken unter großen Beschleunigungen. BAM Bericht 1–3, 10.Juni.

[2] Baeßler, Matthias 2005. BAM-Vorhaben 7229 Setzungsproblematik auf Brücken – DB Projekt Neubewertung der Schotterbeschleunigung auf Brücken. BAM Bericht, 24.01.

[3] Kruse, Holger 2002. Modellgestützte Untersuchung der Gleisdynamik und des Verhaltens von Eisenbahnschotter. VDI Fortschritt Berichte, Reihe 12, Nr. 508, Juni.

[4] Reinecke, J.-M.; Wächter, W. 1994. Querfestigkeit der Holz- und Betonschwellengleise im destabilisierten und stabilisierten Zustand. Deutsche Bahn AG, VersA-Bericht, Nr. 356001, 11.10.

[5] Walter, E.; Uelsmann, U. 2000. Messtechnische Untersuchung an zwei WiB-Überbauten über die Goethestraße in Köln-Lövenich zur Analyse der Resonanzgefährdung. Bericht 00/21, DB-Netz AG, 25.08.

[6] Walter, E.; Uelsmann, U. 2001. Messtechnische Untersuchung an einem WiB-Überbau über die Havel in Fürstenberg zur Analyse der Resonanzgefährdung. Bericht 01/04, DB-Netz AG, 30.03.

[7] Walter, E.; Uelsmann, U. 2001. Messtechnische Untersuchung an einem WiB-Überbau über die Steinförder Straße in Fürstenberg zur Analyse der Resonanzgefährdung. Bericht 01/05, DB-Netz AG, 05.04.

[8] Walter, E.; Uelsmann, U. 2001. Resonanzuntersuchungen an Rahmentragwerken der ABS 4/S13 Köln-Düren. Bericht 03/35, DB-Netz AG, Nov.

[9] Walter, E.; Uelsmann, U. 2003. Verschiebungs- und Beschleunigungsmessungen an der EÜ Burgwiesenstraße. Bericht 03/76, DB-Netz AG, 15.12.

[10] Walter, E.; Uelsmann, U. 2003. Verschiebungs- und Beschleunigungsmessungen an der EÜ über die Landstraße L 264. Bericht 03/77, DB-Netz AG, 15.12.

[11] Zacher, M. 2002. 2. Ergänzung zu Gutachten 902-GU-05-02. DB Systemtechnik Sachverständigenorganisation, 14.06.

CHAPTER 8

Dynamic analysis of structures under high speed train loads: case studies in Spain

F. Gabaldón Castillo, F. Riquelme, J.M. Goicolea-Ruigómez & J.J. Arribas
Escuela de Ingenieros de Caminos, Technical University of Madrid, Spain

ABSTRACT: A relevant aspect in the design of structures under high speed train loads is the dynamical effects associated to the moving loads of the train. In this paper some calculation methods that are usual in the design of high speed train structures are considered. These methods are based on the integration of the motion equations, with or without considering the vehicle-structure interaction. Some applications of the vehicle-structure interaction models are shown in order to analyse the dynamical response of short span bridges, with emphasis in the possible reduction of the dynamical effects associated to these models. To this end the results obtained using both interaction models and moving point loads are compared. Finally some dynamic analyses of singular bridges in the Spanish high speed lines are presented, showing results of interest for the design of these viaducts. The methodology followed in these calculations corresponds to the Spanish IAPF 2006 code and the Eurocode 1 (EN1991-2).

1 INTRODUCTION

Nowadays in Spain the main part of the investment in transport infrastructure is dedicated to funding the construction of new railway lines for high speed trains, being this item the most important too in other countries as France. These new railway lines are a very competitive alternative to other transport modes for medium distances. At this moment in Spain there are three high speed railways in operation: Madrid–Sevilla, Córdoba–Málaga and Madrid–Lérida, pertaining the last one to the railway line Madrid–Barcelona–France. The ADIF authority has in project or construction several railway lines as Madrid–Valencia–Murcia, Madrid–Segovia–Valladolid, Valladolid–Santiago–Porto, Madrid–Badajoz–Lisboa, Sevilla–Huelva–Faro, etc. being the three later considered in the frame of the international high speed railway system Portugal–Spain–Rest of Europe.

These activities have remarked out an important engineering aspect joined specifically to the design of high speed railway structures: the dynamical effects associated to the train moving loads, for which basic solutions have been described by Timoshenko and Young [14] and discussed fully by Fryba [6, 7].

Most engineering design codes for railway bridges have followed the approach of the dynamic factor proposed by UIC [15], which takes into account the dynamic effect of a single moving load and yields a maximum dynamic increment of $\varphi' = 32\%$ for an ideal track without irregularities. The irregularities have been token into account with another parameter φ'', leading to the dynamic factor $\Phi = 1 + \varphi' + \varphi''$. This approach cover the dynamic effects associated to a single moving load but does not include the possibility of resonant effects due to the periodicity of the moving loads, as this phenomenon does not appear for train speeds below 200 km/h. For velocities upper than 200 or 220 km/h and distances of axles between 13 m. and 20 m. resonant effects may appear. An illustrative example showing experimental resonant measurements and the corresponding computational results is shown in [2] for the Spanish AVE train crossing at 219 km/h, the Tagus bridge.

New European codes include the need for dynamic calculations covering resonant behaviour [4, 5, 10]. Generally the calculation procedures are based in the direct integration of the structural response for moving loads modelling the axles of the train. These methods can be applied following several methodologies: the structural model may be analysed either by the complete discrete system with N degrees of freedom and a time integrator like the Newmark method, or by a prior modal analysis and a posteriori time integration of the n significant eigenmodes. Besides, these models can take into account the vehicle–bridge interaction. If vehicle–bridge interaction is considered the complexity of the model is increased and a major computational effort is required. Although these kind of simulation are very interesting from a research point of view they are not useful for standard design calculations except for unusual situations.

2 MODELS BASED ON DIRECT INTEGRATION WITH MOVING POINT LOADS

These class of methods are based on the direct time integration of the dynamic equations of the structure, under the actions corresponding to a train of moving loads of fixed values which values are representative of each axle of the train. The structural model may be analysed through the integration of the complete discrete system with N degrees of freedom, or through a reduction of the number of degrees of freedom via a previous modal analysis which reduces substantially the number of equations. This modal analysis can be performed through an approximate numerical procedure in order to obtain de eigenfrequencies and eigenmodes of vibration. This kind of procedures are available in the majority of finite element codes, and for certain cases of simple structures the spectral analysis can be achieved by analytical methods.

Finite element methods are applicable to arbitrary structures, even when non linear effects must be considered. A spatial semidiscretisation of the structure is performed into subdomains called "finite elements" leading a discrete N-d.o.f. system of equations [17]:

$$\mathbf{M}\ddot{\boldsymbol{d}} + \mathbf{C}\dot{\boldsymbol{d}} + \mathbf{K}\boldsymbol{d} = \boldsymbol{f}(t) \tag{1}$$

where $\mathbf{M}$ is the mass matrix, $\mathbf{C}$ is the damping matrix, $\mathbf{K}$ is the stiffness matrix, $\boldsymbol{f}(t)$ is the vector of external loads, and $\boldsymbol{d}$ is the unknown vector of nodal displacements. In order to integrate in time this system of equations, a direct integration of the model solves the complete system (1) for each time step, and because the equations are generally coupled they must be solved simultaneously. This procedure is valid even when nonlinear effects must be taken into account. In such case the elastic internal forces and viscous damping should be replaced by a general nonlinear term $\boldsymbol{f}^{int}(\boldsymbol{d}, \dot{\boldsymbol{d}})$.

If as usual the structural behaviour is linear, a modal analysis can be performed [1] resulting in another system with a remarkable reduction of degrees of freedom. In a first stage, the eigenvalue problem corresponding to the undamped system is performed solving the generalised eigenproblem corresponding to the structural discrete system:

$$(-\omega^2\mathbf{M} + \mathbf{K})\mathbf{a} = 0 \tag{2}$$

obtaining the more significant n eigenfrequencies ω_i $i = 1, \ldots, n$, and the corresponding normal vibration modes $\boldsymbol{a}_n$, being generally $n \ll N$. In a second stage, the vibration modes $\boldsymbol{a}_i$ oscillating with their respective frequencies ω_i are integrated in time. With this procedure the equations are decouple, and the modal response of each mode is obtained from the dynamic equation of a system with a single degree of freedom.

The simplest procedure to define the train loads is applying load histories in each node. For time step t_i and an axle load F, a nodal load F_J is assigned to the node J if the axle is above an element that contains node J. The magnitude of F_J depends linearly on the distance from the axle to the node. This procedure is outlined in Figure 1 for a single load. This scheme has been implemented in the finite element program FEAP [13] both for the ten HSLM–A (High Speed Load Model) trains

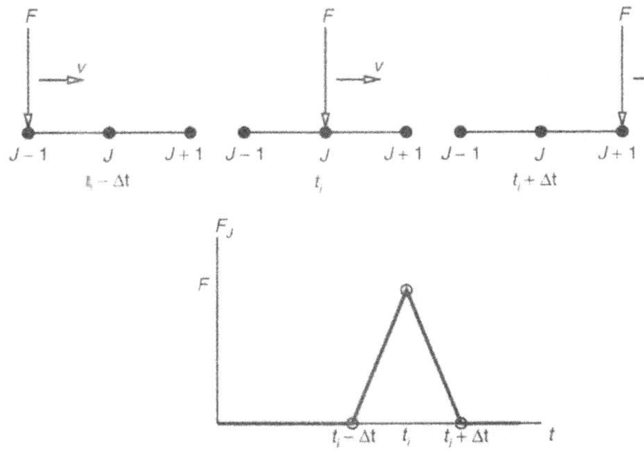

Figure 1. Nodal force time history definition for an axle load F moving at velocity v.

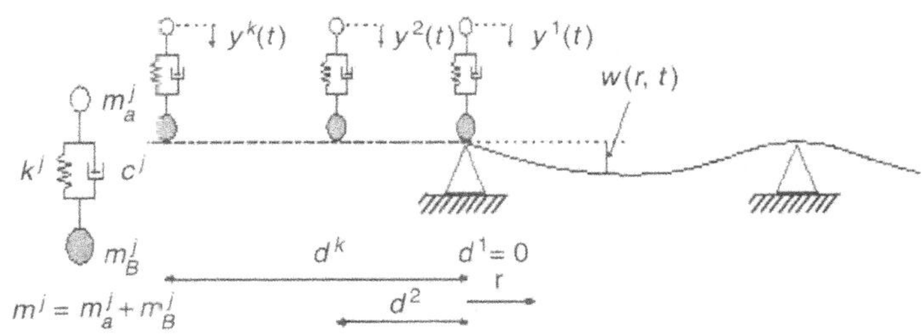

Figure 2. Load train with vehicle–bridge interaction: simplified interaction model and definition of the variables.

defined in the Eurocode [4, 10] and for the seven real trains defined in the Spanish code IAPF [10]. All the results for moving point loads described in this paper have been obtained with the methodology described in this section, using a direct time integration of the significant eigenmodes.

3 MODELS CONSIDERING VEHICLE-BRIDGE INTERACTION

When the structural model considers the relative vertical movement of the vehicles, the dynamic behaviour of the structure is represented more realistically than using moving point loads. With these models the train is considered as a set of masses representing the non-suspended masses of the bogies, the suspended masses of the coaches, springs modelling the suspension of the vehicles, dampers, etc [16]. In this work the simplified interaction models shown in Figure 2 are considered [8]. For a train with k axles, each axle j is represented with an interaction element j with a non-suspended mass m_s^j, a suspended mass m_a^j, and stiffness and dampers constants with values k^j and c^j, respectively.

The equations obtained for the structure and the vehicle are (3) and (4), respectively:

$$\begin{pmatrix} \mathbf{M}_{\mathrm{WW}} & \mathbf{M}_{\mathrm{WB}} \\ \mathbf{M}_{\mathrm{BW}} & \mathbf{M}^{\mathrm{W}}_{\mathrm{BB}} \end{pmatrix} \begin{Bmatrix} \ddot{\boldsymbol{w}} \\ \ddot{\boldsymbol{b}} \end{Bmatrix} + \begin{pmatrix} \mathbf{C}_{\mathrm{WW}} & \mathbf{C}_{\mathrm{WB}} \\ \mathbf{C}_{\mathrm{BW}} & \mathbf{C}^{\mathrm{W}}_{\mathrm{BB}} \end{pmatrix} \begin{Bmatrix} \dot{\boldsymbol{w}} \\ \dot{\boldsymbol{b}} \end{Bmatrix} + \begin{pmatrix} \mathbf{K}_{\mathrm{WW}} & \mathbf{K}_{\mathrm{WB}} \\ \mathbf{K}_{\mathrm{BW}} & \mathbf{K}^{\mathrm{W}}_{\mathrm{BB}} \end{pmatrix} \begin{Bmatrix} \boldsymbol{w} \\ \boldsymbol{b} \end{Bmatrix} = \begin{Bmatrix} \boldsymbol{f}_{\mathrm{W}} \\ \boldsymbol{f}_{\mathrm{B}}^{\mathrm{w,ar}} + \boldsymbol{f}_{\mathrm{B}}^{\mathrm{w,ext}} \end{Bmatrix} \tag{3}$$

$$\begin{pmatrix} \mathbf{M}_{\mathrm{BB}}^{\mathrm{y}} & \mathbf{M}_{\mathrm{BY}} \\ \mathbf{M}_{\mathrm{YB}} & \mathbf{M}_{\mathrm{YY}} \end{pmatrix} \begin{Bmatrix} \ddot{\boldsymbol{b}} \\ \ddot{\boldsymbol{y}} \end{Bmatrix} + \begin{pmatrix} \mathbf{C}_{\mathrm{BB}}^{\mathrm{y}} & \mathbf{C}_{\mathrm{BY}} \\ \mathbf{C}_{\mathrm{YB}} & \mathbf{C}_{\mathrm{YY}} \end{pmatrix} \begin{Bmatrix} \dot{\boldsymbol{b}} \\ \dot{\boldsymbol{y}} \end{Bmatrix} + \begin{pmatrix} \mathbf{K}_{\mathrm{BB}}^{\mathrm{y}} & \mathbf{K}_{\mathrm{BY}} \\ \mathbf{K}_{\mathrm{YB}} & \mathbf{K}_{\mathrm{YY}} \end{pmatrix} \begin{Bmatrix} \boldsymbol{b} \\ \boldsymbol{y} \end{Bmatrix} = \begin{Bmatrix} \boldsymbol{f}_{\mathrm{B}}^{\mathrm{y,ar}} + \boldsymbol{f}_{\mathrm{B}}^{\mathrm{y,ext}} \\ f_{\mathrm{Y}} \end{Bmatrix} \tag{4}$$

In these equations the degrees of freedom have been got into three groups corresponding to the nodes modelling the contact of the train with the deck ($\boldsymbol{b}$), the rest of the nodes of the deck ($\boldsymbol{w}$), and those corresponding to the suspended masses of the vehicle ($\boldsymbol{y}$). In the load vector modelling the wheel-deck forces, the contribution of the action–reaction term ($\boldsymbol{f}^{\cdot,\mathrm{ar}}$) and the external loads have been distinguished ($\boldsymbol{f}^{\cdot,\mathrm{ext}}$).

3.1 *Time integration based on the modal decomposition*

The solution of the equation (3) can be expressed:

$$\begin{Bmatrix} \boldsymbol{w} \\ \boldsymbol{b} \end{Bmatrix} (x_k(t), t) = \sum_{i=1}^{N} q_i(t) \begin{Bmatrix} \phi_{W_i}(x_k(t)) \\ \phi_{B_i}(x_k(t)) \end{Bmatrix} \tag{5}$$

being $q_i = q_i(t)$ the modal amplitude, N the number of modes used in the analysis, ϕ_{W_i} and ϕ_{B_i} the components of the eigenvector i corresponding to the degrees of freedom $\boldsymbol{w}$ and $\boldsymbol{b}$ respectively, and $x_k(t)$ the parameter of position of the axle k.

Normalising the eigenvectors with the mass matrix, the modal equations are:

$$\ddot{q}_i + 2\xi_i\omega_i\dot{q}_i + \omega_i^2 q_i = \begin{Bmatrix} \phi_{\mathrm{W}_i} \\ \phi_{\mathrm{B}_i} \end{Bmatrix}^T \begin{Bmatrix} f_{\mathrm{W}} \\ \boldsymbol{f}_{\mathrm{B}}^{\mathrm{W,ar}} + \boldsymbol{f}_{\mathrm{B}}^{\mathrm{W,ext}} \end{Bmatrix}, \quad i = 1, \ldots, N \tag{6}$$

Solving for $\boldsymbol{f}_{\mathrm{B}}^{\mathrm{y,ar}}$ in (4) and substituting $\boldsymbol{f}_{\mathrm{B}}^{\mathrm{W,ar}} = -\boldsymbol{f}_{\mathrm{B}}^{\mathrm{y,ar}}$, results:

$$\boldsymbol{f}_{\mathrm{B}}^{\mathrm{w,far}} = \boldsymbol{f}_{\mathrm{B}}^{\mathrm{y,ext}} - \mathbf{M}_{\mathrm{BY}}\ddot{\boldsymbol{y}} - \mathbf{M}_{\mathrm{BB}}^{\mathrm{y}}\ddot{\boldsymbol{b}} - \mathbf{C}_{\mathrm{BY}}\dot{\boldsymbol{y}} - \mathbf{C}_{\mathrm{BB}}^{\mathrm{y}}\dot{\boldsymbol{b}} - \mathbf{K}_{\mathrm{BY}}\boldsymbol{y} - \mathbf{K}_{\mathrm{BB}}^{\mathrm{y}}\boldsymbol{b} \tag{7}$$

The equation (6) is re-written via the substitution of the modal expression of $\boldsymbol{b}$ detailed in (5), in equation (7):

$$\begin{aligned} \ddot{q}_i + 2x_i\omega_i\dot{q} + \omega^2 q_i &= \phi_{\mathrm{W}_i} f_{\mathrm{W}} + \phi_{\mathrm{B}_i} f_{\mathrm{B}}^{\mathrm{ext}} - \phi_{\mathrm{B}_i}\mathbf{M}_{\mathrm{BY}}\ddot{\boldsymbol{y}} - \phi_{\mathrm{B}_i}\mathbf{C}_{\mathrm{BY}}\dot{\boldsymbol{y}} - \phi_{\mathrm{B}_i}\mathbf{K}_{\mathrm{BY}}\boldsymbol{y} \\ &\quad - \phi_{\mathrm{B}_i}\mathbf{M}_{\mathrm{BB}}^{\mathrm{y}} \sum_{j=1}^{N} \{\ddot{q}_j\phi_{\mathrm{B}_j} + 2v\dot{q}_j\phi'_{\mathrm{B}_j} + v^2 q_j\phi''_{\mathrm{B}_j}\} \\ &\quad - \phi_{\mathrm{B}_i}\mathbf{C}_{\mathrm{BB}}^{\mathrm{y}} \sum_{j=1}^{N} \{\dot{q}_j\phi_{\mathrm{B}_j} + v q_j\phi'_{\mathrm{B}_j}\} \\ &\quad - \phi_{\mathrm{B}_i}\mathbf{K}_{\mathrm{BB}}^{\mathrm{y}} \sum_{j=1}^{N} \{q_j\phi_{\mathrm{B}_j}\} \end{aligned} \tag{8}$$

being: $\boldsymbol{f}_{\mathrm{B}}^{\mathrm{ext}} = \boldsymbol{f}_{\mathrm{B}}^{\mathrm{x,ext}} + \boldsymbol{f}_{\mathrm{B}}^{\mathrm{y,ext}}$ y $x_k = vt$ (with the hypothesis of constant velocity of the train).

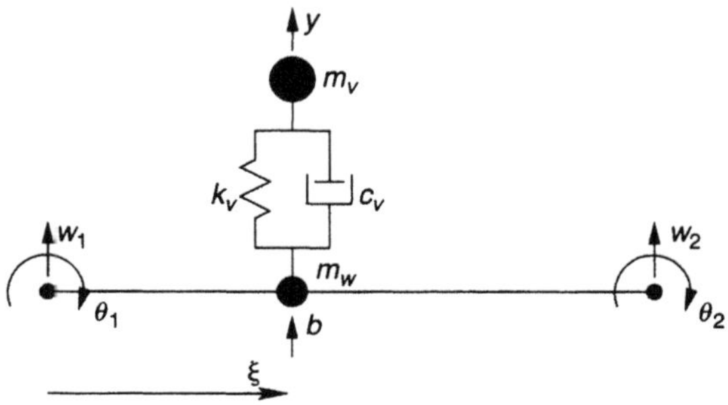

Figure 3. Bernoulli beam element with simplified vehicle–bridge interaction model.

In order to integrate the movement of the suspended masses the modal expression of $\boldsymbol{b}$ is substituted in (4), resulting the system of differential equations:

$$\begin{aligned}\mathbf{M}_{\mathrm{YY}}\ddot{\boldsymbol{y}} + \mathbf{C}_{\mathrm{YY}}\dot{\boldsymbol{y}} + \mathbf{K}_{\mathrm{YY}}\boldsymbol{y} = \mathbf{f}_{\mathrm{Y}} &- \mathbf{M}_{\mathrm{YB}}\sum_{j=1}^{N}\{\ddot{q}\phi_{\mathrm{B}_j} + 2v\dot{q}_j\phi'_{\mathrm{B}_j} + v^2 q_j\phi''_{\mathrm{B}_j}\}\\ &- \mathbf{C}_{\mathrm{YB}}\sum_{j=1}^{N}\{\dot{q}_j\phi_{\mathrm{B}_j} + vq_j\phi'_{\mathrm{B}_j}\}\\ &- \mathbf{K}_{\mathrm{YB}}\sum_{j=1}^{N}\{q_j\phi_{\mathrm{B}_j}\}\end{aligned} \tag{9}$$

The equations (8) and (9) must be integrated in order to obtain the modal amplitudes $q_j(t)$ and the degrees of freedom of the suspended masses $\boldsymbol{y}(t)$. After this, the degrees of freedom corresponding to the movement of the deck ($\boldsymbol{w}$ and $\boldsymbol{b}$) are computed from equation (5).

3.2 *Direct time integration*

In order to carry out the time integration of the whole structure, taking into account all its degrees of freedom, a model based on the Euler–Bernoulli beam with an additional degree of freedom corresponding to the axle of the train was developed (see Figure 3). Let w_1, θ_1, w_2, θ_2 denote the degrees of freedom of the beam (displacement and rotation of the nodes), and y and b the degrees of freedom of the suspended mass (m_v) and the non-suspended one (m_w) respectively. The dynamical equilibrium equations of the element are:

$$\mathbf{M}_{\mathrm{interac}}\begin{Bmatrix}\ddot{y}\\ \ddot{b}\end{Bmatrix} + \mathbf{C}_{\mathrm{interac}}\begin{Bmatrix}\dot{y}\\ \dot{b}\end{Bmatrix} + \mathbf{K}_{\mathrm{interac}}\begin{Bmatrix}y\\ b\end{Bmatrix} = \begin{Bmatrix}F_y\\ F_b\end{Bmatrix} \tag{10}$$

being:

$$\mathbf{M}_{\mathrm{interac}} = \begin{pmatrix}m_v & 0\\ 0 & m_w\end{pmatrix},\quad \mathbf{C}_{\mathrm{interac}} = \begin{pmatrix}c_v & -c_v\\ -c_v & c_v\end{pmatrix},\quad \mathbf{K}_{\mathrm{interac}} = \begin{pmatrix}k_v & -k_v\\ -k_v & k_v\end{pmatrix} \tag{11}$$

The displacement of the non-suspended mass b is interpolated from the nodal displacements of the beam in a standard way via the hermitic shape functions [17]:

$$b = [N_1 \quad N_2 \quad N_3 \quad N_4] \begin{Bmatrix} w_1 \\ \theta_1 \\ w_2 \\ \theta_2 \end{Bmatrix} \tag{12}$$

Using the transformation matrix **T**:

$$\mathbf{T} = \begin{pmatrix} 0 & 0 & 1 & 0 & 0 \\ N_1 & N_2 & 0 & N_3 & N_4 \end{pmatrix} \tag{13}$$

The motion equations of the element result [9]:

$$\begin{Bmatrix} F_1 \\ M_1 \\ F_y \\ F_2 \\ M_2 \end{Bmatrix} = \mathbf{M} \begin{Bmatrix} \ddot{w}_1 \\ \ddot{\theta}_1 \\ \ddot{y} \\ \ddot{w}_2 \\ \ddot{\theta}_2 \end{Bmatrix} + \mathbf{C} \begin{Bmatrix} \dot{w}_1 \\ \dot{\theta}_1 \\ \dot{y} \\ \dot{w}_2 \\ \dot{\theta}_2 \end{Bmatrix} + \mathbf{K} \begin{Bmatrix} w_1 \\ \theta_1 \\ y \\ w_2 \\ \theta_2 \end{Bmatrix} \tag{14}$$

being:

$$\mathbf{M} = \mathbf{T}^T \mathbf{M}_{\text{interac}} \mathbf{T}, \ \mathbf{C} = \mathbf{T}^T \mathbf{C}_{\text{interac}} \mathbf{T}, \ \mathbf{K} = \mathbf{T}^T \mathbf{K}_{\text{interac}} \mathbf{T} \tag{15}$$

The matrix (15) must be assembled with the standard matrix of the Bernoulli beam element in order to obtain the stiffness matrix of the interaction element. The element matrix must be re-computed in each time step.

4 APPLICATION: DYNAMIC ANALYSES OF SHORT SPAN BRIDGES

In these analyses four short span bridges defined in the ERRI 214 catalogue of isostatic bridges have been considered. The definition of the bridges is summarised in Table 1. The loads considered was the corresponding to the ICE 350 E high speed train, that is running in Spain with the name AVE 103, in the range of velocities from 120 km/h to 420 km/h. The same damping ratio $\xi = 2\%$ has been adopted for all the calculations.

The curves in Figures 4 and 5 correspond to the envelopes of accelerations and displacements in the mid point of the 7.5 m span. It can be observed a left translation of the peaks computed with the interaction model relative to the point load results. This effect appears because the interaction models include the mass of the vehicle leading to the reduction of the frequencies of the viaduct (if both masses have similar values). Besides, a reduction in the highest values of accelerations and displacement is obtained when using interaction models.

Table 1. Definition of short span bridges from ERRI D214 catalogue.

Length (m)	Mass (kg $\times 10^3$/m)	Frequency (Hz)	EI (kNm^2)
5.0	7	16	453919
7.5	9	12	1661921
10.0	10	8	2593823
20.0	20	4	50660592

With the standard finite element formulation of the Euler–Bernoulli beam element [17] the expression of the terms in the mass matrix is:

$$M_{ij} = \rho I \int_L N_{i,x} N_{j,x} dx + \rho A \int_L N_i N_j dx \tag{16}$$

The first term in the right hand side of equation (16) corresponds to the rotational part of the mass matrix ($\mathbf{M}_{rot}$), and the second term contributes to the translational part ($\mathbf{M}_{tras}$), being $\mathbf{M} = \mathbf{M}_{rot} + \mathbf{M}_{tras}$. The contribution of $\mathbf{M}_{rot}$ to $\mathbf{M}$ is very small for long span bridges. Nevertheless, although this term is neglected by several computational codes, it is important in short span bridges. This fact is shown in Table 2 where the values of the first frequency of an isostatic beam are reported for several lengths.

Figure 6 shows the comparison between the envelopes of displacements for the spans of $L = 5.0$ m and $L = 20.0$ m. It has been considered models with rotational mass and without it, using both point loads as vehicle-bridge interaction elements.

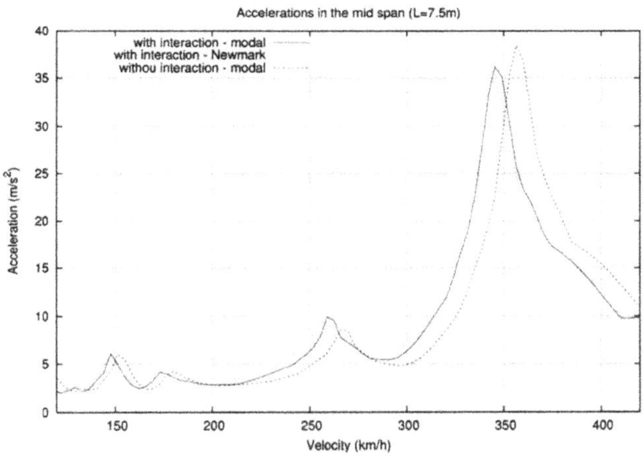

Figure 4. Acceleration and displacement in the mid span of the 7.5 m beam.

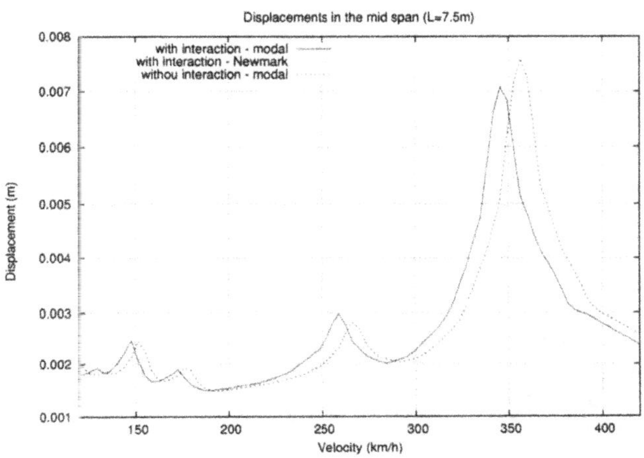

Figure 5. Acceleration and displacement in the mid span of the 7.5 m beam.

Table 2. First frequency of an isostatic beam considering/not considering the rotational mass matrix.

	Frequency (Hz)	
Length (m)	With R.M.	Without R.M.
5.0	13.5478	16.0000
7.5	11.0682	12.0000
10.0	7.6322	8.0000
20.0	3.9515	4.0000

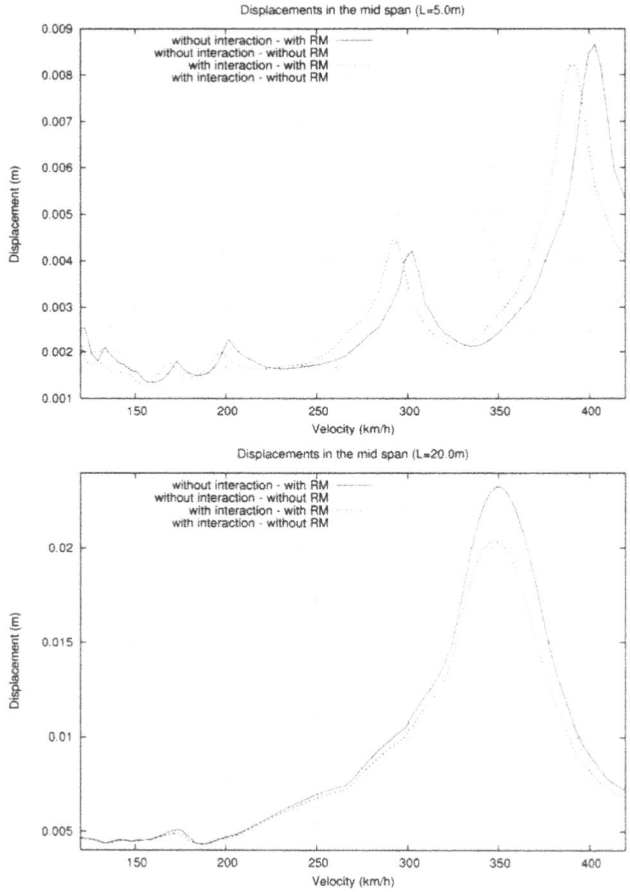

Figure 6. Vertical displacement in the midspan: $L = 5$ and $L = 20$ m.

For the 5 m span bridge, comparing the results obtained with rotational mass (black and blue curves) with those obtained without it (red and yellow curves) an important translation to the left side is observed for the models with rotational mass. The reason is the remarkable increment of the structural mass associated to the rotational terms. Comparing, for the same formulation of the mass matrix, the results obtained using moving point loads with those obtained with interaction models (the black curve versus the blue one, and the red curve versus the yellow one) another translation of the resonant peaks is observed, although less relevant than in the former comparison.

Finally, for the 20 m span can be concluded that the contribution of the rotational is negligible from a practical point of view.

5 DYNAMIC ANALYSIS OF AN ARCH BRIDGE

5.1 *Description of the structure and modelling*

In this section the results obtained from the dynamic analysis of an arch bridge, designed for the Ourense–Santiago high speed line, are shown. This viaduct is formed by a concrete arch with a span of 170 m and a deck with length 630 m. The deck and the arch are connected with five piles. The material used for the arch and deck is concrete H40 (characteristic resistance $f_{ck} = 40\,\text{kN/mm}^2$) and for the piles concrete H30 ($f_{ck} = 30\,\text{kN/mm}^2$). The area and moment of inertia of the arch are $7.76\,\text{m}^2$ and $14.87\,\text{m}^4$, respectively. For the deck these values are $A = 11.25\,\text{m}^2$ and $I = 25.91\,\text{m}^4$.

The dynamic analysis was carried out with a 2D finite element model. The model have 1639 beam elements, leading to a system with 4872 degrees of freedom. All the computations have been carried out with the finite element code iris [12]. The mesh of the model is showed in Figure 7. The joins of the piles with the deck are encastred, having the pile in this point the same rotation that the deck. In the left end support the horizontal displacement and bending rotation are allowed, and the right end support is a fixed point.

For the time integration the first 114 eigenmodes have been considered in order to take into account all the frequencies lower than 30 Hz, and a modal damping $\xi = 2\%$ has been assigned to all the modes. Figure 8 shows the four lower eigenmodes computed.

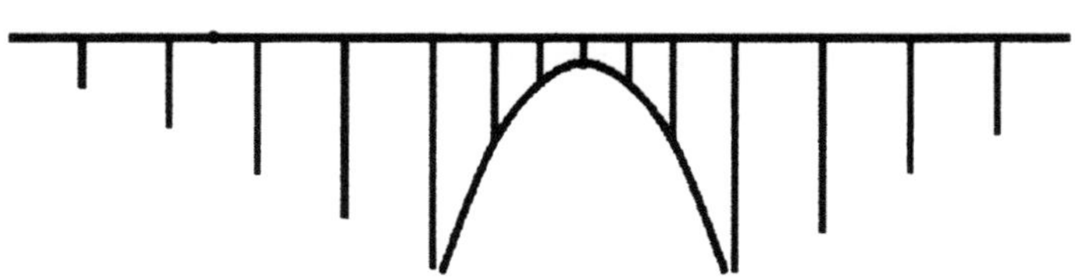

Figure 7. Arch bridge. Finite element model.

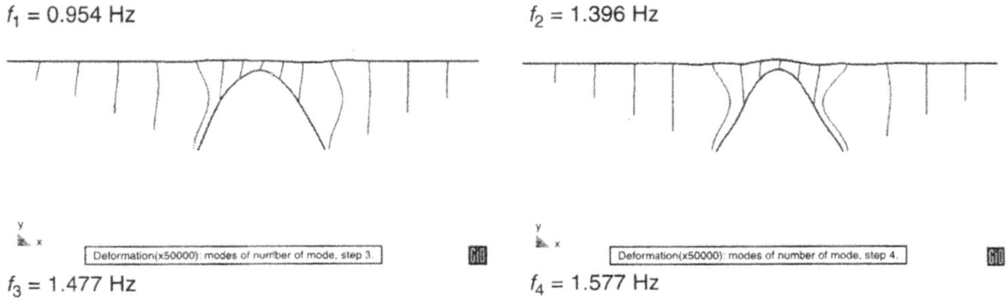

Figure 8. Arch bridge. First to fourth eigenmodes.

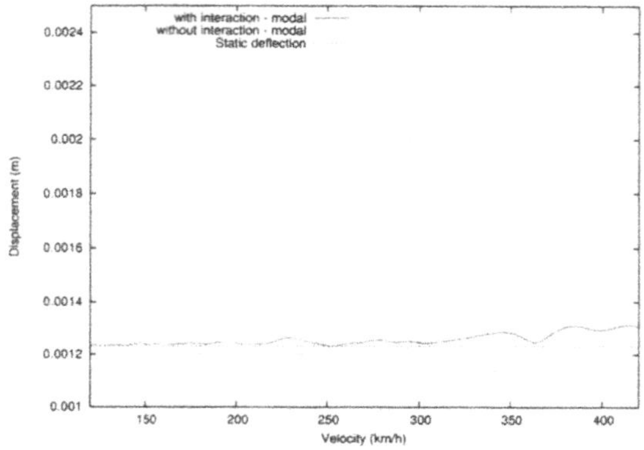

Figure 9. Arch bridge. Displacements in the 3rd midspan.

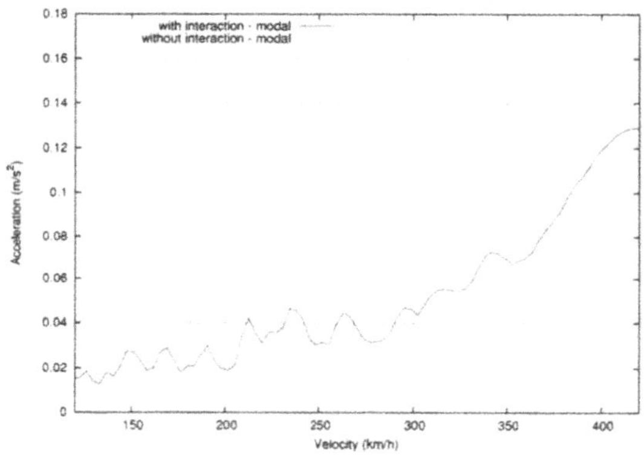

Figure 10. Arch bridge. Accelerations in the 3rd midspan.

5.2 *Results*

The calculations were performed for a range of velocities from 120 to 420 km/h every $\Delta v = 5$ km/h, considering the following high speed trains:

1. The ten trains corresponding to the HSLM-A (High Speed Load Model) IAPF 2006 [10] and Eurocode [4].
2. The seven European high speed trains defined in the Spanish IAPF code [10]: AVE, EUROSTAR 373/1, ETR-Y, ICE-2, TALGO-AV, THALYS and VIRGIN, modelled with moving point loads.
3. The AVE 103 Spanish high speed train modelled both with moving point loads as with simplified interaction elements, in order to compare the results obtained with these methodologies.

Figures 9 and 10 show the displacements and acceleration in the centre of the 3rd span obtained with the AVE-103 train (with and without interaction). Figures 9 and 10 show these variables in the keystone of the arch.

It can be observed from these figures that the dynamical effects are more relevant in the keystone of the arch. Besides there isn't reduction of the dynamical effects with the interaction models because the spans of the structure have moderately long lengths.

Figures 13 and 14 show the bending moments in the left support and in the keystone of the arch respectively, computed both for the seven real trains defined in the Spanish code IAPF [10] and for the ten HSLM-A (High Speed Load Model) trains defined in the Eurocode [4, 10]. By a direct

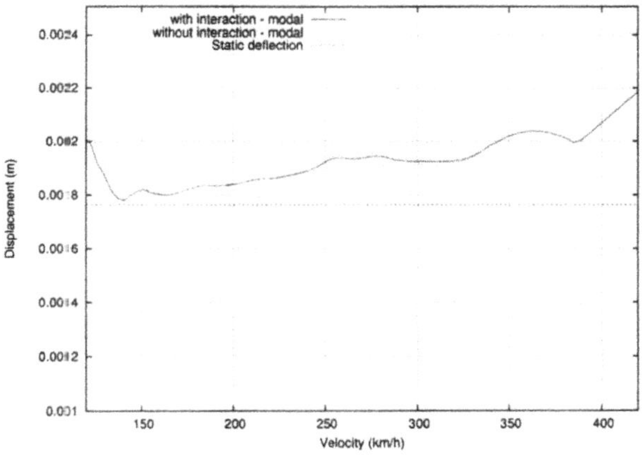

Figure 11. Arch bridge. Displacements in the keystone of the arch.

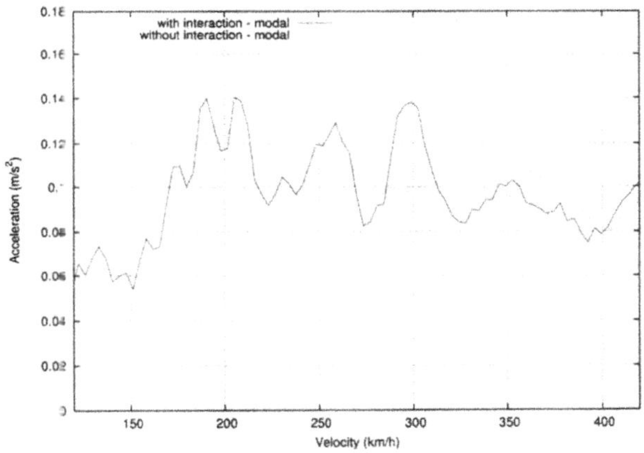

Figure 12. Arch bridge. Accelerations in the keystone of the arch.

comparison of the highest dynamical values with the static envelopes in each point the dynamic amplification factor obtained is ≈2.5.

5.3 *Conclusions*

The two principals eigenmodes mainly contain the deformation of the arch. Their frequencies are lower than those usual in standard bridges with shorter lengths. The first mode ($f_1 = 0.954$ Hz) leads to non symmetric movements and the second mode ($f_2 = 1.396$ Hz) is symmetric. The following modes correspond to the movement of the deck and the piles, and in consequence is convenient to take into account the dynamical effects in these elements.

Resonant effects could appear for low velocities due to the low values of the main frequencies. Nevertheless, because the wave length of the first modes (≈170 m) is one order of magnitude higher than the length of the train coaches (between 13 and 27 m) there are simultaneously several axle loads in the bridge and therefore their effects are mutually cancelled.

Finally, in this structure the consideration of interaction model is not relevant because the results obtained are very similar with those computed with moving point loads.

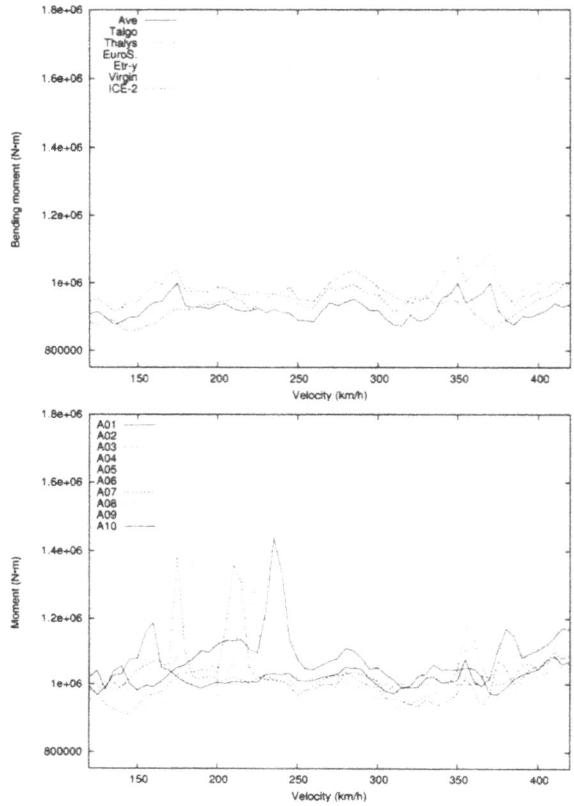

Figure 13. Arch bridge. Bending moments in the left support of the arch.

6 DYNAMIC ANALYSIS OF A PERGOLA BRIDGE

6.1 *Description of the structure and modelling*

This viaduct will be in the Madrid–Valencia high speed line. It is a pergola bridge composed of 64 attached longitudinal beams with box section typology. The length of the precasted beams varies from 35.35 m of the shortest one to 40.62 m. The separation between the axles of the beams is 4.5 or 6 m, leading to 326.75 m of total width.

In order to verify the requirements in the Spanish IAPF code [10] and Eurocode 1 [4], the dynamic analysis comprises the following points:

1. Determination of the impact coefficient for velocities lower than 220 km/h, via simplified methods.
2. Determination of the impact coefficient for velocities higher than 220 km/h, via dynamic calculations. These computations consider the following aspects:
   - Time integration of all the eigenmodes with frequencies lower than 30 Hz.
   - The actions applied are the corresponding to the ten trains of the HSLM-A model (High Speed Load Model A) defined in [4, 10], which are dynamic envelopes of the effects of the possible real trains.
   - The calculations are performed for a range of velocities from 120 to 300 km/h every $\Delta v = 5$ km/h. The highest velocity is 20% higher than the design velocity $v_{design} = 250$ km/h.

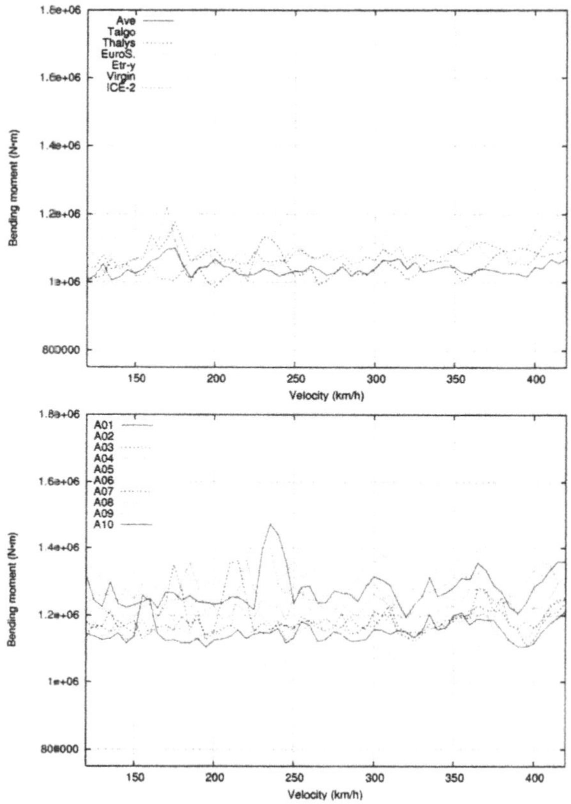

Figure 14. Arch bridge. Bending moments in the keystone of the arch.

3. Verification of the ELS (Usage Limit State) requirements for the maximum values of acceleration, vertical displacements and rotations

Taking into account the dimensions and typology of the viaduct, and because the tracks are mainly in transversal direction (see Fig. 15), it has been considered convenient to perform a three dimensional finite element model.

This bridge has been modelled with 7220 four nodes shell elements and 6342 nodes leading to a model with 37730 degrees of freedom. Figures 16 and 17 show a general view of the mesh. The boundary conditions adopted correspond to a point support of the ends of each beam as is showed in the detail of Figure 18. For the evaluation of the vibrating mass of the viaduct it has been considered the mass of concrete and the added masses corresponding to other non-structural members (ballast, sleepers, rails, etc). The values adopted for the mechanical properties of concrete are Young modulus $E = 3.5$ GPa and Poison ratio $\nu = 0.2$.

6.2 *Dynamic analysis of the structure*

The dynamic analysis of the structure has been performed in accord to the requirements in the appendix B of IAPF 2006 code [10], via the time integration of the modes of oscillation of the structure. In order to take into account all the modes with associated eigenfrequencies lower than 30 Hz, it is necessary to consider 552 eigenmodes in the dynamic computations. The deformed shapes corresponding to the fourth lower modes of oscillation are showed in figure 19. The damping

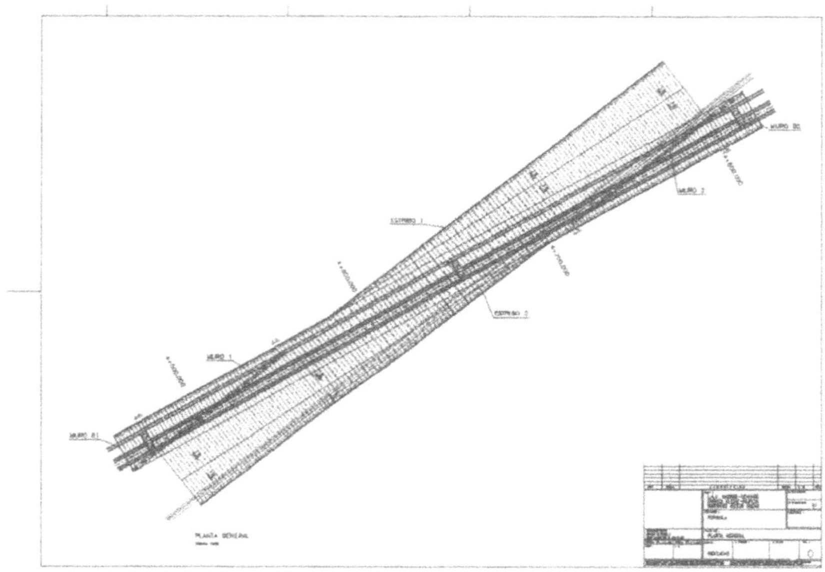

Figure 15. Pergola bridge. General plant view.

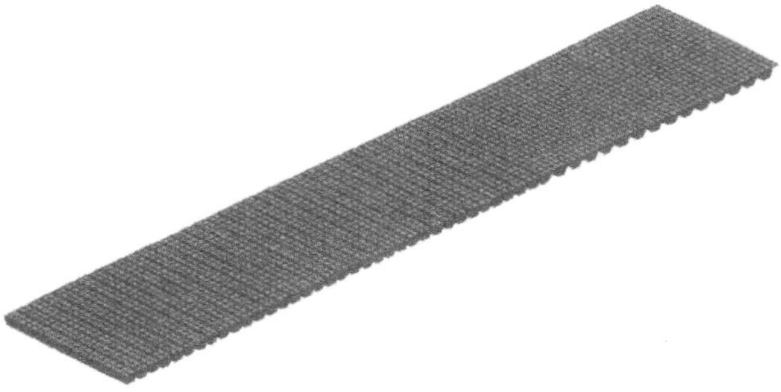

Figure 16. Finite element model of the pergola bridge. Lower view.

Figure 17. Finite element model of the pergola bridge. Upper view.

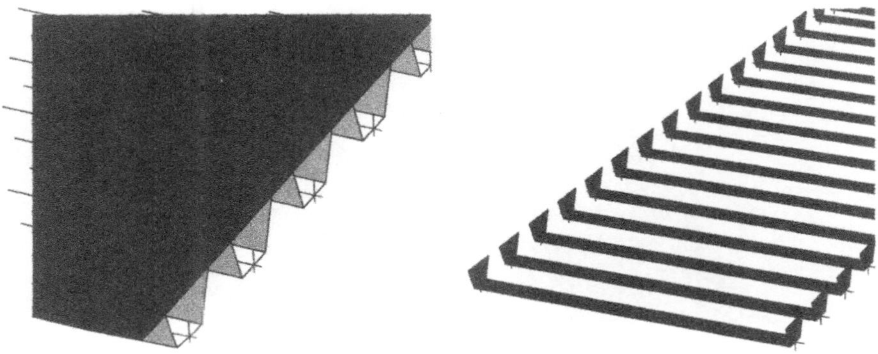

Figure 18. Finite element model of the pergola bridge. Detail of the model.

is defined as Rayleigh damping [1] considering a fraction of critical damping rate ξ equal to 2% (for concrete structures). The applied loads correspond to the 10 trains defined in the HSLM-A model for a range of velocities from 120 km/h to 300 km/h every 5 km/h, leading to 370 analyses with 552 eigenmodes each one. To perform this computations in a reasonable lapse of time they have been used 24 PENTIUM machines, with 2.6 GHz and 512 Mb RAM, running in parallel processes.

The results obtained from each one of the 370 analyses have been post-processed in order to obtain the maximum values of vertical displacement, vertical acceleration and "in plane" rotations. These maximum values have been considered for reporting the results obtained in the following sections.

6.3 *Simplified calculation of the impact coefficient*

The simplified methodology for the evaluation of the impact coefficient Φ_2 described in the IAPF 2006 code [10] is only valid for velocities lower than 220 km/h, and has the following requirements:

1. The typology of the viaduct must be conventional.
2. The first eigenfrequency must be valued between the two bounds defined in the expression (2.5) of the IAPF 2006 code [10].

For the particular case of this viaduct the lower bound is $f_{min} = 2.73$ Hz and the upper one $f_{max} = 6.23$ Hz. Therefore the first frequency of the viaduct ($f_1 = 4.31$ Hz) verifies this requirement. For structures with good maintenance the expression of the simplified impact coefficient is:

$$\Phi_2 = \frac{1.44}{\sqrt{L_\Phi - 0.2}} + 0.82 = 1.06 \tag{17}$$

being L_Φ defined in the IAPF 2006 [10] as the "characteristic length" of the viaduct (for this bridge $L_\Phi = 38$ m).

6.4 *Calculation of the impact coefficient from dynamic analysis*

The Spanish IAPF 2006 code requires a complete and specific dynamic analysis of structures in high speed lines with design velocity upper than 220 km/h. From the results of this analysis using the high speed trains, an impact coefficient Φ must be obtained. This impact coefficient Φ will be compared with the impact coefficient Φ_2 from expression (17), and the coefficient with the highest value will be considered for the design of the structure.

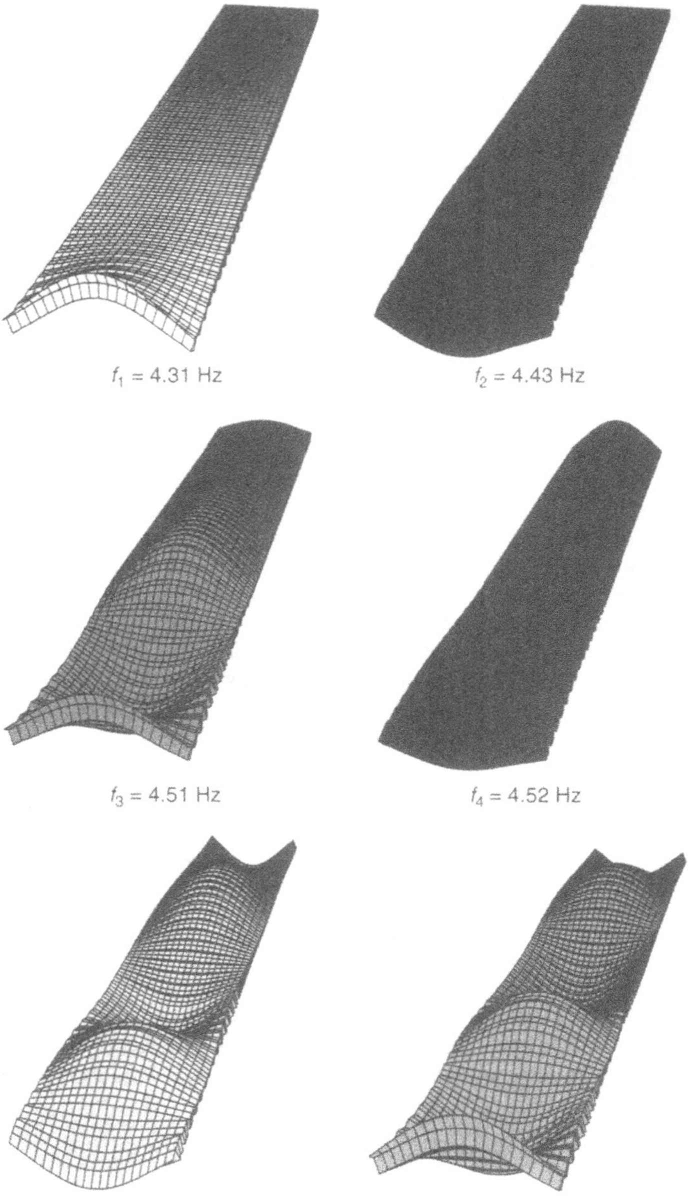

Figure 19. Pergola bridge. First to fourth eigenmodes.

From the dynamic analysis results, the real impact coefficient of the train i in some point of the structures is:

$$\Phi_i = (1 + 0.5\varphi'')\frac{\delta^i_{\text{din,rel}}}{\delta_{\text{est,tipo}}} \tag{18}$$

being $\delta^i_{\text{din,rel}}$ the maximum movement obtained for the train i, $\delta_{\text{est,tipo}}$ the movement obtained statically using the train defined in the IAPF 2006 ("tren tipo") from the UIC train, and φ'' a coefficient related to the tracking irregularities [3] (in this bridge $\varphi'' = 0.0141$).

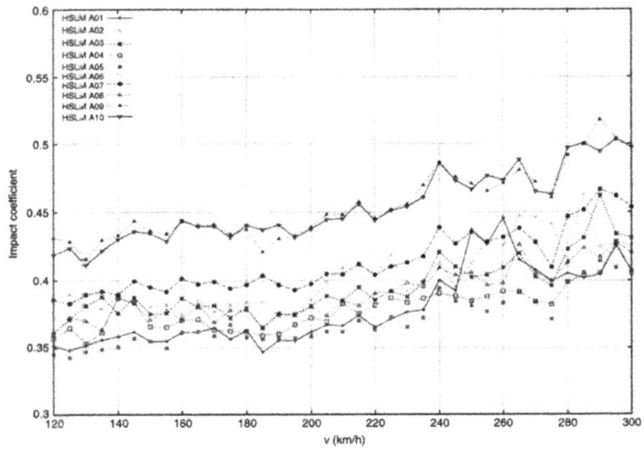

Figure 20. Pergola bridge. Impact coefficient Φ in the midpoint of the track.

Figure 20 shows the impact coefficient Φ obtained applying the expression (18) for each velocity and each train.

The impact coefficient is the highest value obtained considering all the trains, and in this structure is the one obtained with the HSLM-A9 train running with a velocity $v = 290$ km/h:

$$\Phi = \max_i \Phi_i = 0.517 \tag{19}$$

Comparing this value of Φ with the one obtained in (17) using the simplified methodology results the impact coefficient to consider in the design of the viaduct:

$$\Phi = 1.06 \tag{20}$$

6.5 *Verification of the usage limit state*

The deformations, displacements and accelerations in the deck of railway bridges are related to the Usage Limit States for the structure, and to the Ultimate Limit State of the track and vehicles [11]. For this reason in the IAPF 2006 and Eurocode [10, 4] are included the Usage Limit States that must be considered in order to avoid security problems.

6.5.1 *Accelerations*

In accord to the section 4.2.1.1.1 of IAPF 2006 and section A.2.4.4.2.1 of Eurocode 1, the maximum vertical acceleration of the deck must verify:

$$a_{\max} \leq 0.35\,g \tag{21}$$

for bridges with ballast. In accord to [3], the values obtained from the dynamic computations (for an ideal track) must be incremented by a factor $(1 + 0.5\varphi'')$ in order to take into account the track irregularities. The maximum values of the acceleration are showed in Table 3. From these results can be concluded that the structure verifies the requirement (21).

6.5.2 *Rotations of the deck*

For bridges with double track and ballast the values of the rotation in the transition tie bar-deck are bounded with the value [10, 4]:

$$\theta \leq 3.5 \cdot 10^{-3}\,\text{rad} \tag{22}$$

Table 3. Pergola bridge. Maximum values of the vertical acceleration.

Train	a (m/s^2)	a (m/s^2)
HSLM A01	1.89 (v = 250 km/h)	2.01 (v = 260 km/h)
HSLM A02	2.41 (v = 265 km/h)	3.00 (v = 285 km/h)
HSLM A03	1.89 (v = 290 km/h)	2.19 (v = 300 km/h)
HSLM A04	2.97 (v = 295 km/h)	2.98 (v = 300 km/h)
HSLM A05	1.92 (v = 290 km/h)	1.79 (v = 295 km/h)
HSLM A06	1.95 (v = 285 km/h)	1.79 (v = 290 km/h)
HSLM A07	2.16 (v = 290 km/h)	1.87 (v = 295 km/h)
HSLM A08	1.52 (v = 265 km/h)	2.57 (v = 295 km/h)
HSLM A09	1.95 (v = 290 km/h)	1.98 (v = 295 km/h)
HSLM A10	2.06 (v = 285 km/h)	2.22 (v = 295 km/h)

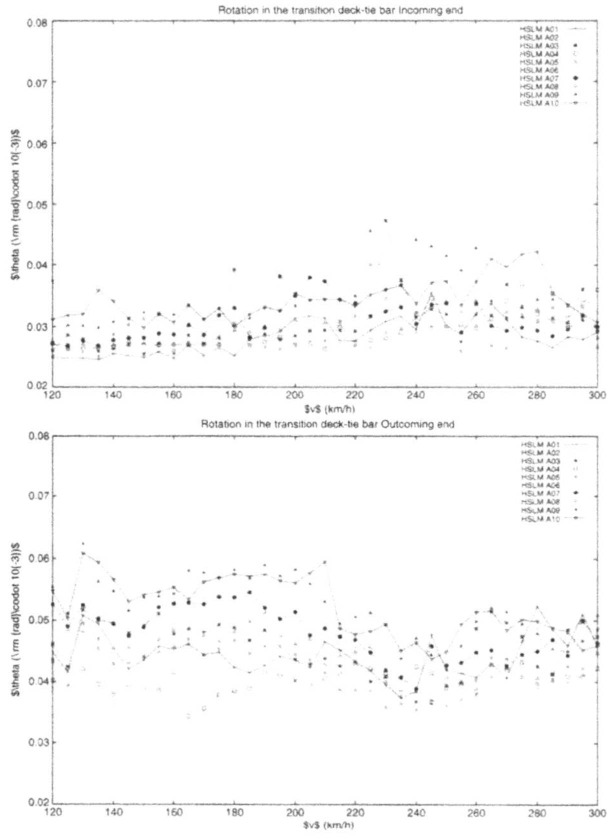

Figure 21. Pergola bridge. Envelopes of rotation in the deck-tie bar transition.

Figure 21 shows the envelopes of the rotations both in the incoming as the out-coming tie bars of the bridge. It can be verified that the computed values are two orders of magnitude lower than the required one.

6.5.3 *Comfort*

In section 4.2.1.2 of the IAPF 2006 code [10] the values of the vertical displacements are limited in order to ensure the comfort of the passengers. Besides the comfort level is classified in accord to the maximum values obtained for the vertical acceleration. In this bridge the maximum acceleration has a value upper than 2.0 m/s^2, and therefore the comfort level is considered as "allowable".

The requirements for comfort are established in terms of the ratio L/δ, being L the length and δ the maximum vertical displacement. The value of this ratio must be corrected with a factor that depends on the maximum acceleration value and the number of spans. For the Pergola bridge described in this paper the comfort requirements are satisfied.

ACKNOWLEDGEMENTS

The authors acknowledge the financial support of Ministerio de Fomento of Spain to the project "Análisis dinámico de estructuras sometidas a acciones de trenes de alta velocidad" through the research program "Acciones Estratégicas del Área Sectorial de Construcción Civil y Conservación del Patrimonio Histórico Cultural" of the "Plan Nacional de Investigación Científica, Desarrollo e Innovación Tecnológica 2002–2003".

REFERENCES

[1] Clough, R.W. & Penzien, J. 1994. *Dynamics of Structures*. Second Edition, McGraw-Hill.
[2] Domínguez Barbero, J. 2001. *Dinámica de puentes de ferrocarril para alta velocidad: métodos de cálculo y estudio de la resonancia*. Ph. D. thesis (in Spanish). E.T.S. Ingenieros de Caminos, Universidad Politécnica de Madrid.
[3] ERRI D214 committee. 1999. *Ponts-rails pour vitesses >200 km/h; Etude Numérique de l'influence des irrégularités de voie dans les cas de résonance des ponts*. Rapport technique RP 5. European Rail Research Institute (ERRI). March.
[4] Eurocode 1. 2003. *Actions on structures – Part 2: Traffic loads on bridges*. CEN.
[5] Ferrovie dello Stato, Italy. 1997. *Sovraccarichi per il calcolo dei ponti ferroviari. Istruzioni per la progettazione, l'esecuzione e il collaudo*. Code I/SC/PS-OM/2298.
[6] Fryba, L. 1972. *Vibration of solids and structures under moving loads*. Academia, Noordhoff.
[7] Fryba, L. 1996. *Dynamics of railway bridges*. Thomas Telford.
[8] Gabaldón, F., Riquelme, F.Y. & Goicolea, J.M. 2005. *Análisis dinámico de estructuras sometidas a acciones de trenes de alta velocidad, considerando la interacción vehículo-estructura*. Métodos Numéricos en Ingeniería. SEMNI/APMTAC.
[9] Ju, S. & Lin, H. 2003. Numerical Investigation of a steel arch bridge and interaction with high-speed trains. *Engineering Structures*. Vol. 25. pp. 241–250.
[10] Ministerio de Fomento, Spain. 2006. *Instrucción de acciones a considerar en el proyecto de puentes de ferrocarril*. Draft J.
[11] Nasarre & de Goicochea, J. 2002. *Estados límites de servicio en relación con la vía en puentes de ferrocarril. Puentes de Ferrocarril*. Proyecto, Construcción y Mantenimiento. Congreso del Grupo Español de IABSE. Vol. 2. Madrid, Junio.
[12] Romero, I. *iris: User Manual*. Universidad Politécnica de Madrid. http://w3.mecanica.upm.es/~ignacio/iris.html.
[13] Taylor, R.L. *FEAP. A Finite Element Analysis Program*. University of California, Berkeley. http://www.ce.berkeley.edu/~rlt/feap.
[14] Timoshenko, S. & Young, D. 1995. *Vibration problems in engineering*. Van Nostrand, 3rd ed..
[15] Union Internationale des Chemins de Fer. 1979. *Charges a prendre en considerations dans le calcul des ponts-rail*. Code UIC 776-1R.
[16] Yang, Y.B., Yau, J.D. & Wu, Y.S. 2004. *Vehicle-Bridge Interaction Dynamics With Applications to High-Speed Railways*, World Scientific.
[17] Zienkiewicz, O.C. & Taylor, R.L. 2000. *The finite element method*, Butterworth-Heinemann.

CHAPTER 9

Dynamic calculations of high-speed railway bridges in France – some case studies

W. Hoorpah
MIO, Paris, France

ABSTRACT: New HSR lines are now under construction in France. Dynamic analysis under the high speed trains is systematically carried by the contractor's design office in the final design stage. This is required systematically for steel and steel concrete composite decks which are now built for all the major bridges. In the last HSR line, TGV Est some concrete bridges with complex geometry are now also checked for their dynamic behaviour. The dynamic analysis design is now commonly practiced with commercial software by most design offices. The dynamic response is obtained under moving loads of constant amplitudes. Moving vehicle loads with dynamic interaction methods are not used in the final design calculations. This paper presents the dynamic design calculations of HSR bridge types in France with some case studies of particular bridge types: – tied arch bridges, – half through steel girder bridges, – composite twin box and plate girder deck, – concrete portal frame. The general dynamic design strategy and the practical calculation methods will be explained. Some conclusions on the dynamic behaviour of the different deck structures will be pointed out.

1 HIGH SPEED RAILWAY BRIDGE TYPES

1.1 *Historical review*

High speed railway bridges are generally characterised by a stiff deck in order to ensure a satisfactory dynamic behaviour under the train loads crossing at high speeds. In the first HSR lines only concrete decks were supposed to guarantee the necessary stiffness.

Since the first HSR TGV Sud Est line in 1983, the bridge types in the medium and long span ranges (above 40 m) have relentlessly undergone an evolution towards steel composite decks.

While this first line had only pre-stressed concrete box girders, the subsequent lines witnessed the arrival of steel in the deck structure.

The percentage of steel concrete composite decks was about 30% in the TGV Nord in 1993 and about 65% in 1999 for the TGV Méditerrannée. For the TGV East line under construction now this percentage has reached 100%. The twin plate girder composite deck alone accounts for nearly two thirds of the bridges.

One common feature of all these bridges is the concrete deck slab that plays an important role in the durability of the bridge in addition to providing the necessary stiffness for a good dynamic behaviour.

This evolution was possible thanks to the extensive dynamic analysis carried both at the conceptual design and final design phases for the more flexible composite steel concrete decks.

1.2 *Bridge types*

The structural and architectural form of the bridges have been chosen on economic and aesthetic grounds. The structural type depends generally on the span length, the vertical clearance underneath and the skew.

Continuous decks are always chosen. However the length between joints are limited to about 450 m and intermediate isostatic spans are necessary because of track expansion devices limitations. The width of the deck of the HSR bridges carrying two tracks is above 12 m. Somme narrow one track bridge about 6 m wide have been built at line interchange and over-pass.

The most common structure is the twin girder with span up to 75 m for the Moselle crossing in the TGV Est. Half through decks with lateral plate girders are built for motorway crossings with low clearance. For skew crossings, four plate girders or twin box girder composite decks are generally chosen. This type has been chosen for a long span crossing also at Jaulny in the TGV Est.

For long spans above 80 m, tied arch bridges have been built in the TGV Méditerrannée up to 124 m. Before that in the TGV Nord a closed steel truss 90 m span bridge was built over La Deule canal.

For small spans below 30 m, the deck structure is mostly concrete. Over busy roads or railway lines with shallow depth possibility filler beam decks are chosen.

2 DYNAMIC DESIGN STRATEGY

In the initial conceptual design phase, the bridge designer: SNCF or other private design office can now use the Eurocode EN 1991-2 flow chart to determine the need for a complete dynamic analysis or the traditional simplified approach using dynamic amplification factors. For the HSR bridges built up to now, the design method is defined in the SNCF design documents and the special specifications for each line.

2.1 *Dynamic coefficient method for all bridges*

For all bridges the design specifications are given in the SNCF document Fascicule 2.01. Whatever the deck structure the dynamic effects on the structural forces and deformations are due to the following factors:

- the high speed of the moving loads producing inertial forces which are not included in the static calculations;
- the passing of point loads more or les regularly spaced leading to resonance effects;
- the wheel load variations due to the random defects of the tracks.
- the deck frequency;
- the deck damping usually taken at 0.5% of the critical damping for composite bridges.

If the train speed is below 220 km/h and the permanent load deflection is within a specified range these dynamic effects are usually accounted for with the dynamic coefficients Φ_2 applied on the UIC 71 and SW loads.

$$\Phi_2 = \frac{1.44}{\sqrt{L_\Phi} - 0.2} + 0.82 \tag{1}$$

with $1.00 < \Phi_2 < 1.67$

L_Φ the characteristic length depending on the span length and distribution, the deck structural element

For speeds above 220 km/h, a complete dynamic analysis is required under real train convoy Cr.

A new dynamic coefficient Φr is calculated from the ratio of the dynamic deflection over the static deflection.

$$\Phi_r = MAX\left[\frac{y_{dyn}}{y_{sta}}\right] \tag{2}$$

If the amplified effects $Cr \times \Phi_r < UIC\ 71 \times \Phi_2$ then the usual design rules for speeds $V < 220$ Km/h are used.

If Cr × Φ_r > UIC 71 × Φ_2, either the special design rules under the real train convoy are to be used, or the bridge structure is redesigned to satisfy the first condition.

2.2 *Dynamic analysis for HSR bridges*

In the previous HSR lines, the experience of the SNCF led to the dynamic analysis of only composite steel concrete deck bridges. For the recent TGV Est line, almost all the bridges have been checked with a dynamic analysis.

The design data are for the HSR bridges are:

- design speeds:
  - a commercial speed 300 or 350 km/h;
  - and a potential speed of 350 or 400 km/h;
  - 3 critical speed below the commercial speed;
  - a quasi-static speed of 10 km/h.
- a series of specific HSR trains: TGV simple and double train, Talgo, Virgin, ETR460, ETR-Y-500, Thalys 2, Eurostar, ICE 2 and 3. For the last HSR line, Perpignan Figueras, the 10 universal trains of Eurocode 1, A1 to A10 are now used for the dynamic loadings.
- specific limit state design criteria:
  - vertical acceleration of 0.35 g on the rail considering the bridge vibrations for frequencies below 20 Hz for ballasted tracks;
  - end rotation at abutments of 0.002/h radians and rotation of 0.004/h between two consecutive deck (h is the height of the rail over the bearings);
  - torsional rotation of 0.4 mm/m or 1.2 mm for 3.000 m
  - mid span deflection of L/2200 for some TGV Est bridges.

These deformation criteria are to ensure safety regarding

- the contact between the train wheels and the rails
- the decohesion of the ballast

and to ensure passenger comfort by limiting the perceived vehicle acceleration.

3 DYNAMIC CALCULATION METHODS

3.1 *Calculation methods*

Analytical solutions for the partial derivative equations of movement can be used only for simply supported beams under regularly spaced mobile forces of constant amplitude. This case has a theoretical interest to prove the resonance effect due to speed and spacing. However computer programs are mostly used in all instances as this allows a more realistic modelling of the bridge with variable inertia, complex geometry and fast computation under different HSR models. The resulting data can also be processed for further calculations: stresses, fatigue.

The calculation methods with commercial software are very often done with modal superposition methods. A modal analysis of the bridge model is first carried out to obtain the modal frequency values and the modal deformation diagrams.

The frequencies f are related to the critical speeds by the simple formula: Vcr = d × f where d is approximately equal to axle load spacings: TGV, Eurostar, Thalys: 18.7 m; ICE 2: 27.60 m; ICE 3: 24.80 m, ETR 50: 26.20 m; ETR 460: 28.60 m; VIRGIN: 23.90 m, TALGO AV2: 13.14 m and similar values for the 10 Eurocode universal trains.

The dynamic response is then computed for the specified design speeds and the critical speeds.

However the above formula is valid for a regular spacing of identical loads. Due to the irregular pacing of real high speed trains and the different load intensity of the engine wagon, there is a shift in the critical speed giving a slightly higher value of the theoretical spacing of 19 m for the TGV.

Nowadays, it has become usual to carry the dynamic response at by sweeping all the possible speeds between 100 and 400 km/h or higher at 5 km/h interval. This is to ensure that the resonance speed are not left out.

3.2 *Bridge modelling*

For most HSR bridges, only the deck and upper superstructures are modelled. The piers are seldom included as the dynamic behaviour of the deck under moving loads is not dependant on the pier stiffness as in seismic analysis.

The bridge modelling depends on the structural type and the bridge geometry. For most straight twin girder composite bridges a line model made of 3D beam elements is sufficient.

A plane grid or three dimensional model are necessary for half through decks, for curved for decks or complex support arrangement.

Some bridges like the tied arch spans require 3D beam elements combined with shell elements.

3.3 *Load modelling*

The load modelling can be done in different ways. In the first generation HSR with line models, the dynamic loading of the deck structure is applied node by node by a specific pre-processor taking into account the node spacing with an isostatic load distribution at the end nodes of each beam element. (Figure 1).

Modern softwares allows easier ways to describe the convoy passing on special finite elements describing the real trajectory. These elements have no stiffness or mass and they are here only to distribute the loads at each time step to the element nodes. They are powerful enough to calculate particular track alignments and can simulate two trains moving in opposite directions on the deck.

3.4 *Post-processing of dynamic results*

The results of the dynamic calculations: namely bending moments and vertical shear are used to compute the stress spectra from the sectional properties. This is then used to calculate the stress range: normal stress $\Delta\sigma$ and shear stress $\Delta\tau$ by the rain flow counting method. The stress range are then checked against limiting values derived from Eurocode 3 fatigue class details. This is usually done by the Palmgreen-Miner rule by checking that the total calculated damage stays below unit value for the total number of crossings during the life-time of the bridge.

The specific calculation rules for the fatigue verification are:

- double TGV at the commercial speed,
- 70 train passage on the bridge daily,
- 5% of opposite train crossing on the bridge,
- lifetime of one hundred years.

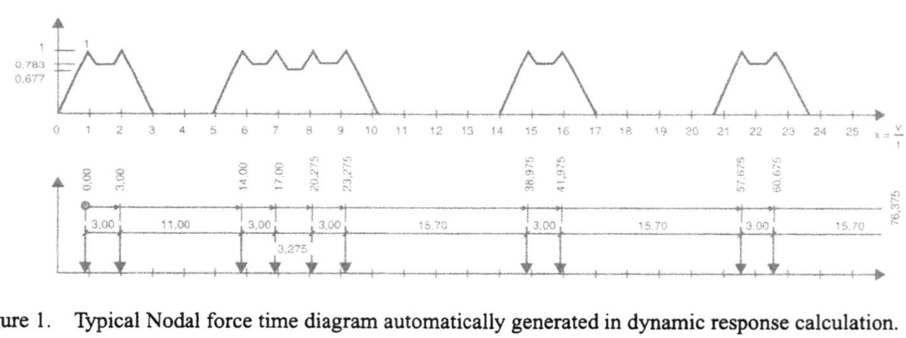

Figure 1. Typical Nodal force time diagram automatically generated in dynamic response calculation.

4 TIED ARCH BRIDGES

4.1 *The dynamic behaviour of some TGV tied arch bridges*

The TGV Méditerranée line had four large tied arch bridges with steel arches and composite decks with different spans and structural design:

- The viaduct of la Garde Adhémar consisting of two continuous tied arches with 110.300 m spans rendered continuous with a central arch and inclined struts. The deck is composed of a composite concrete slab connected to cross beams and stringers. The box section tie girders are on either side of the composite deck.
- The two viaducts of Mornas and Mondragon with spans of 118 m 700 and 82 m 500 respectively having double rib arches linked with vertical rods and wide plate gusset at the crown. The deck is also composed of a composite concrete slab connected to cross beams and stringers with box tie girders.
- The Bonpas tied arch bridge over the A7 toll has the maximum span length of the line: 124 m with steel arch ribs and a composite filler beam deck also connected to the lateral plate tie girders.

Although these bridges are somewhat different in their arch and deck structural design, some general conclusions can be deduced on the dynamic behaviour of tied arch bridges with lateral girders and composite decks.

The design based on static dimensioning gives a sufficiently stiff structure that generally satisfies the dynamic criteria for the bridge structure and for the rail vehicles at the required speeds. This satisfactory behaviour is due to various factors:

- the vertical deformation under one track eccentric loadings causing a large global torsional effect in twin girder decks are absorbed here in double flexural behaviour of each tied arch;
- the low vertical acceleration values can be explained with the low frequency – around 1 Hz for the vertical bending modes of the deck;
- the differential torsional rotation stays below the limiting values as the global behaviour in torsion is quite good.

Table 1 shows the variation of the fundamental bending and torsional frequencies of the four bridges. The longer spans have smaller frequencies due to their lower stiffness. The structural differences can also explain these variations: arch inclination, upper bracings between arches, the relative stiffness of the arches and the tie deck.

4.2 *The Bonpas tied arch bridge over the toll buildings of A7 at South Avignon*

The HSR line crosses the toll entrance to A7 at South Avignon on a viaduct with two single span composite plate girders decks about 30 m long and a central tied arch span of 124 m.

The tied arch deck is composed of two lateral plate girders 2.900 m deep, spaced at 14.5 m between webs and supporting a filler beam deck with transversal HEA 600 rolled beams. The hangers are every 7.500 m. The arch rise is 20 m with box section ribs of size 1.8 m by 2 m with 50 mm plates (Fig. 2).

Table 1. First vertical and torsional modal frequency of ties arch bridges.

	Garde Adhémar	Mornas	Mondragon	Bonpas
Span	110.3 m	118.5 m	82.5 m	124 m
F_{Vert1}	1.31 Hz	1.31 Hz	2.06 Hz	0.96 Hz
$F_{Torsion1}$	2.27 Hz	1.70 Hz	3.07 Hz	1.55 Hz

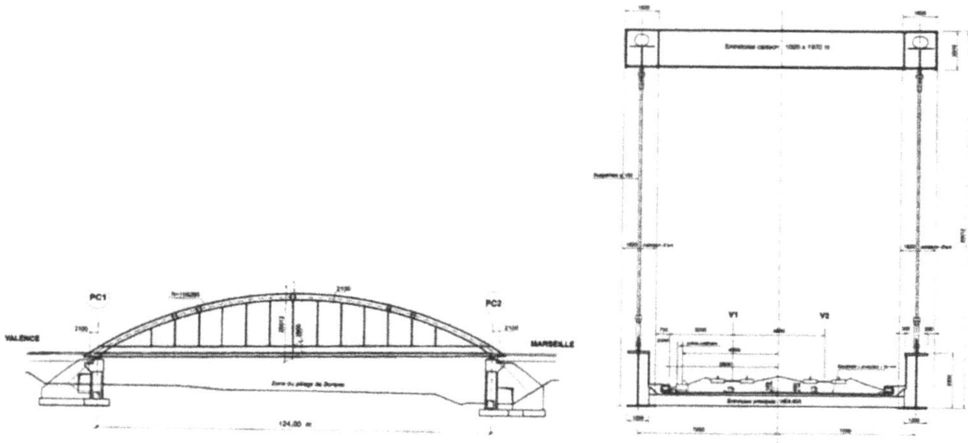

Figure 2. 124 m span tied arch bridge of Bonpas.

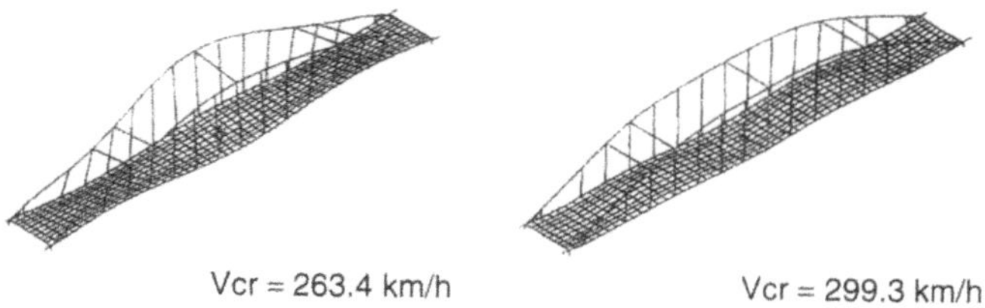

Figure 3. Modal deformations and critical speeds.

The dynamic analysis was carried with a spatial model of the complete structure with 3D beam elements. The pile and abutments are not included in the models; the deck support at the the four corners are represented with appropriate support constraints. The computer model represents:

- the two box section arch ribs,
- the plate girder tie beams,
- the upper bracings between the rach ribs,
- the hangers,
- the filer beams (steel and concrete) between the tie beams,
- longitudinal fictitious stringers to account for the mechanical and mass characteristics of the deck

The arch tie junction are modelled by rigid bars in global bending and shell elements for the local FEM analysis.

This model automatically considers the global torsion due to one track eccentric loading. This torsion causes a differential flexural bending between the two tied arches and gives normal forces in the upper bracings.

The transversal filler beams are grouped by four in the model. The vertical eccentricity between the neutral axes of the cross beams and the longitudinal girders are also added in the input data which classically includes the mechanical and mass properties of the 3D beam elements.

As for the composite deck the cross beams have the mechanical properties of composite sections with a steel/concrete modular ratio $n = 6$ and the fictitious stringers as well as the tie girders a modular ratio of 15.

The dynamic loads are applied on the fictitious stringers nodes.

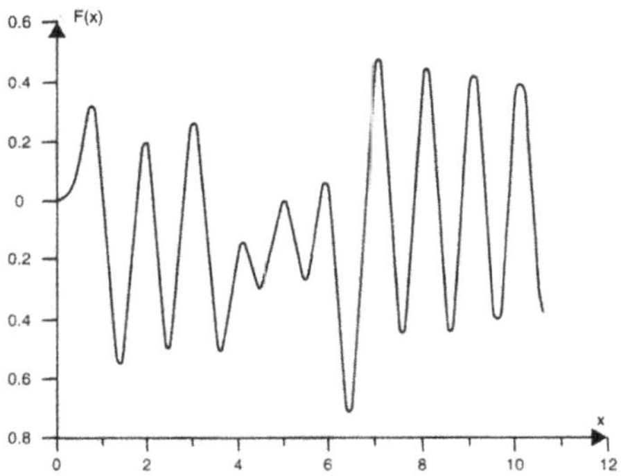

Figure 4a. Quarter span deflection.

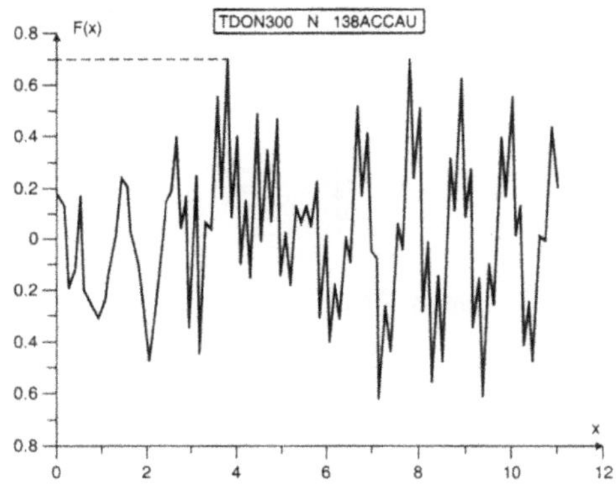

Figure 4b. Quarter span acceleration.

Table 2. Dynamic results at critical and specified speeds.

	152 km/h	225 km/h	300 km/h	350 km/h	Limit
Vertical deflection at ¼ span	10.5 mm	12 mm	15 mm	23.5 mm	
Static deflection = 9 mm Φ_r	1.167	1.333	1.667	2.611	
Vertical acceleration under rail	0.25 mm/s^2	0.34 mm/s^2	0.45 mm/s^2	0.85 mm/s^2	3.43
End rotation of deck	0.51 mrad	0.55 mrad	0.75 mrad	1.15 mrad	1.88
Torsional rotation near abut.	0.04 mm/m	0.041 mm/m	0.050 mm/m	0.049 mm/m	0.4

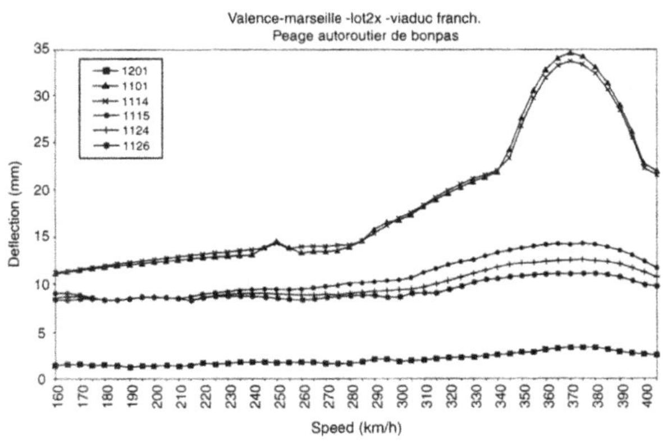

Figure 5. Vertical dynamic deflection with speed sweeping. From 160 to 420 km/h.

5 TWIN BOX GIRDER COMPOSITE BRIDGE: JAULNY VIADUCT IN TGV EST

5.1 *Bridge structure*

The Jaulny Viaduct in the new TGV Est line is a continuous 478.7 m long bridge with the span distribution: 60.50 – 4 × 73.80 – 65.60 – 57.40 m. The fixed pier is P6 with 421 m from the west abutment (the maximum dilatable length of the deck must be below 450 m) and 57.40 m from the East abutment. In this way only the west abutment had a track expansion device. At the East abutment, because of the curved track alignment no expansion device could be installed (Fig. 6). The deck has a curved alignment of radius 6667 m.

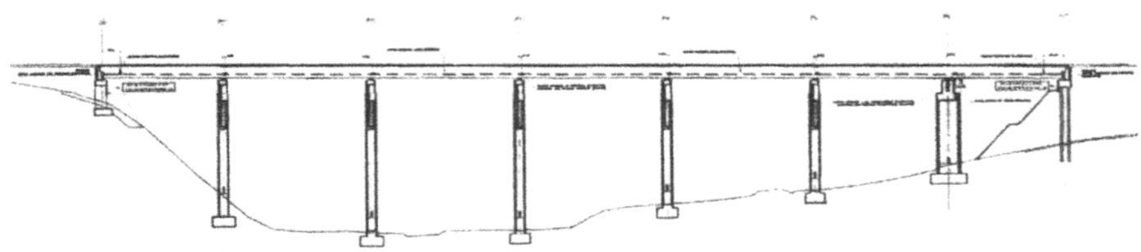

Figure 6. Jaulny Viaduct: Elevation.

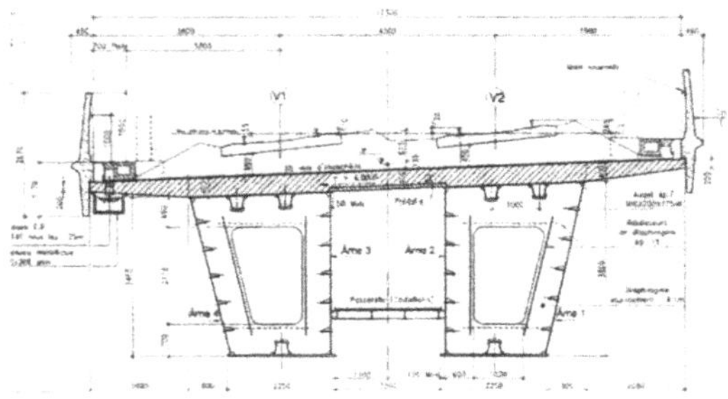

Figure 7. Jaulny Viaduct: Cross section.

Figure 8. Bridge model in FEM software.

The deck is made of two trapezoidal constant depth steel box of external height 3800 mm and internal height 3468 mm. connected to a 40 cm thick reinforced concrete slab. The two box girders are joined by diaphragms on the supports and two or three cross beams only in the spans. The inside of the box sections are not painted and can be visited from the end sections.

5.2 *Dynamic analysis*

Because of the slenderness of the deck structure and the relatively long span, the span distribution was determined by an extensive dynamic analysis from the conceptual design phase.

The imposed design speed were 350 and 400 km/h. The dynamic behaviour of HSR trains mentioned before had to be checked. Because the irregular spacing of the HSR axle loads, the design office used a time step analysis, much more time consuming but supposed to be more precise than the modal superposition method. The time step for the calculation was 1/40th the value of the fundamental bridge period. In this method the damping values have to be given as Rayleigh coefficients α and β times the stiffness and mass matrices respectively computed from two modal damping values.

$$C = \alpha K + \beta M \tag{3}$$

The FE model of the deck was a plane grid of 3D beam.

Figure 9. First vertical bending mode at f = 1.071 Hz.

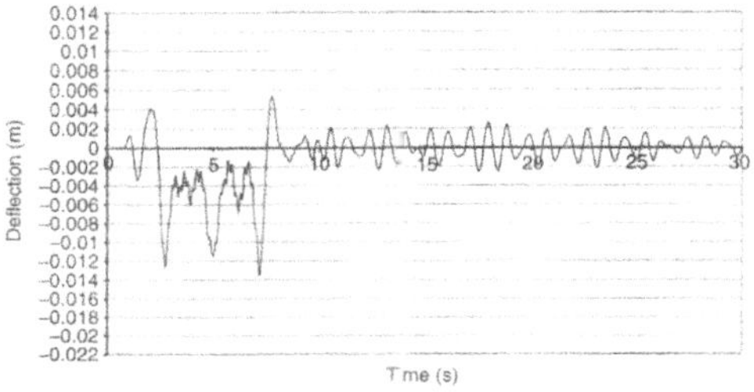

Figure 10a. Mid span deflection under TGV at 350 km/h.

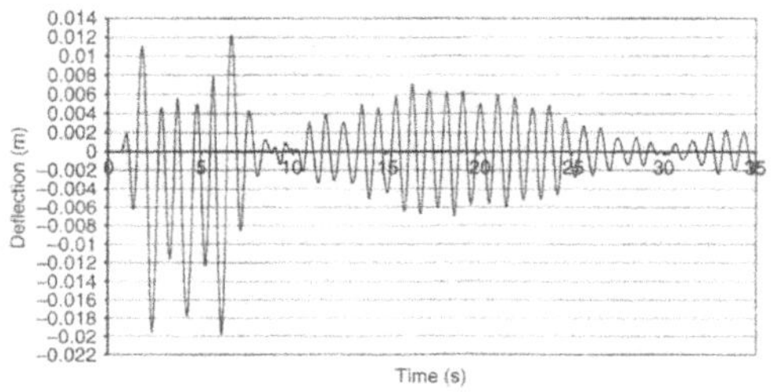

Figure 10b. Mid span deflection under TGV at 410 km/h.

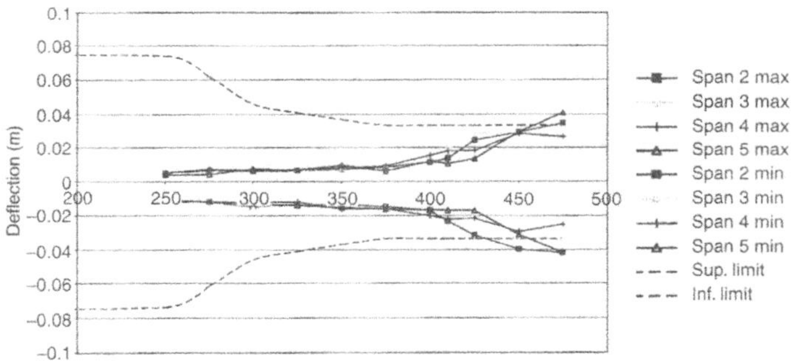

Figure 10c. Midspan deflection between 250 and 475 km/h.

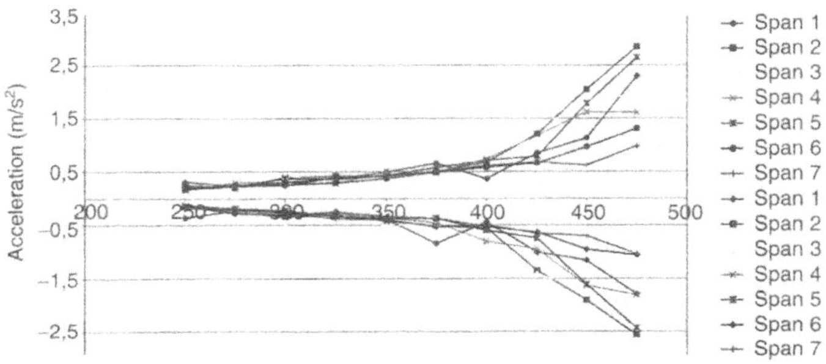

Figure 10d. Midspan acceleration between 250 and 475 km/h.

The critical speed near the imposed design speed values was 393 km/h for the double TGV.

The Figures 10a and 10b show the difference in dynamic behaviour below and above this critical speed.

Mid span deflection and the other dynamic results were computed for speeds between 250 and 475 km/h. As the span distribution is not symmetrical, the two directions of train circulation were studied.

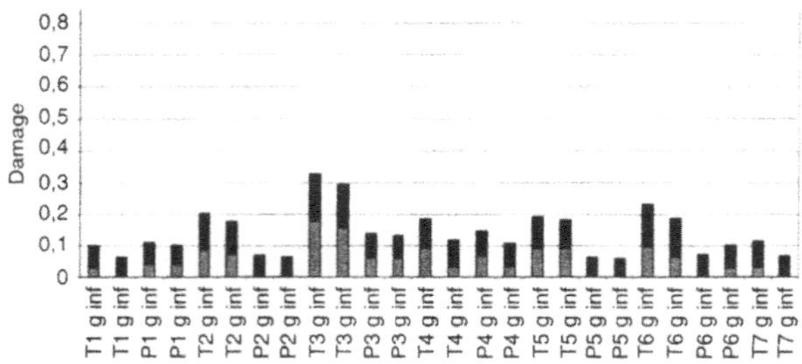

Figure 10e. Fatigue analysis: Damage calculations for lower flange.

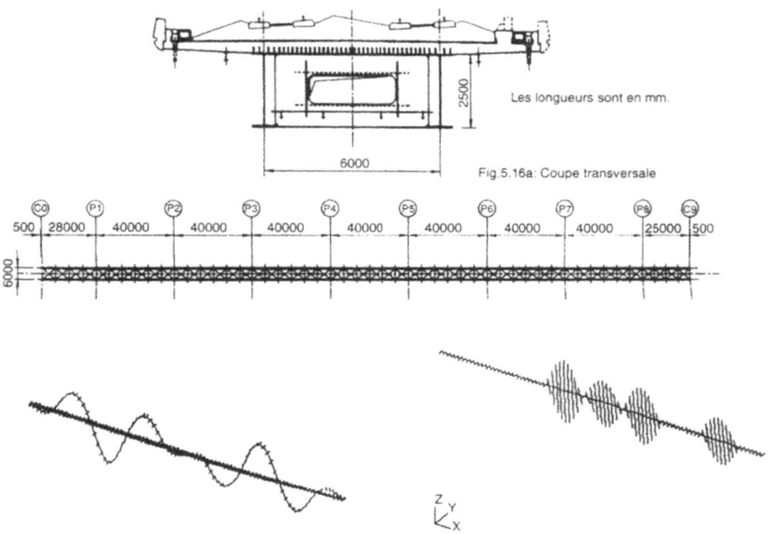

Figure 11. Typical twin plate girder composite bridge with steel lower bracing with bending and torsional modes.

5.3 *Composite twin plate girder deck*

This bridge type is really the most common deck structure both for motorways and HSR. The two girders are connected the upper concrete slab and their lower flanges joined by either a concrete slab or steel bracings. In both cases, the deck behaves as a box section.

The dynamic analysis is carried by a line model with 3D beam elements and transversal needles without mass and stiffness whose only function is to illustrate the torsional modes.

The dynamic loads are applied directly on the model as mobile vertical forces and torques.

Numerous papers have been published on the dynamic analysis of this bridge type.

6 HALF THROUGH BRIDGE OVER A4 AT BUSSY

This highly skew bridge over the A4 motorway for the new TGV Est is half through lateral girder bridge, with transversal composite filler beam. The two span are 49.1 m long each.

The FEM model was a grid with 3D beams. (Figure 12c).

The dynamic analysis was carried for the various European HSR trains (Table 3).

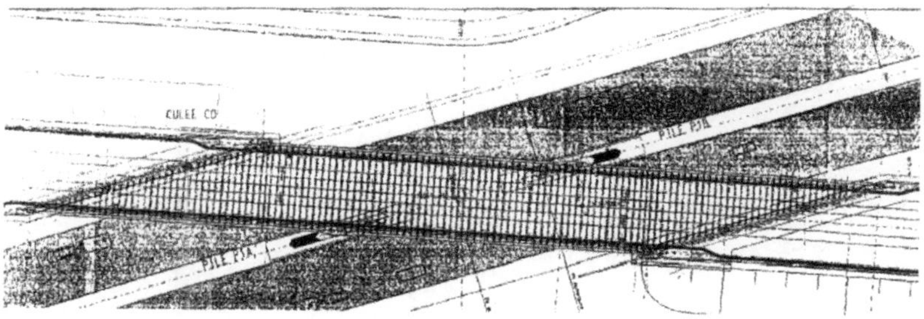

Figure 12a. Plan view.

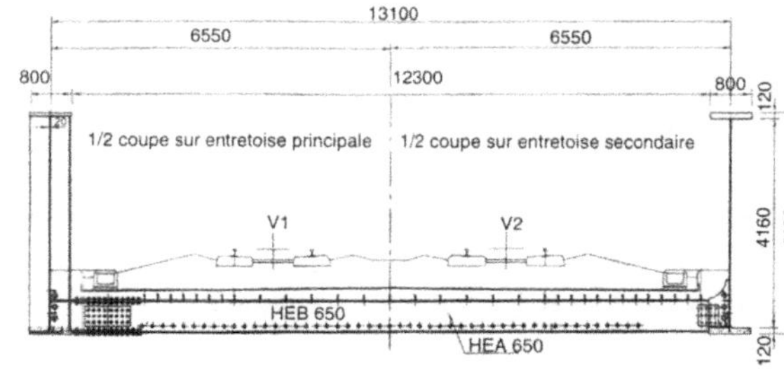

Figure 12b. Cross section.

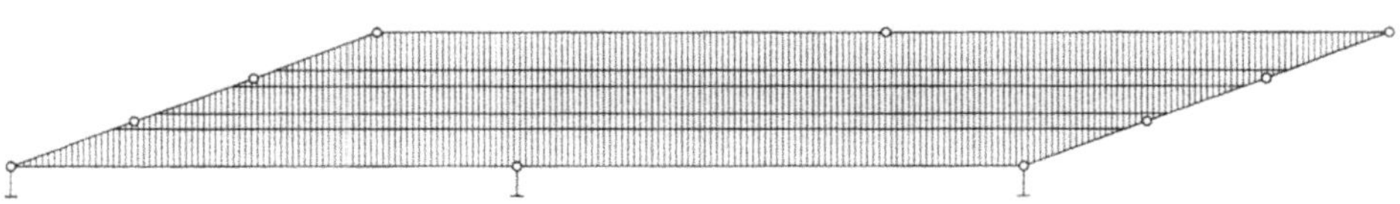

Figure 12c. FE Model with fictitious loading elements.

Table 3. Critical speeds for European trains.

MODE	TYPE OF MODE	Freq (Hz)	V.TGV (km/h)	V.ICE2 (km/h)	V.ICE3 (km/h)	V.ETR Y 500 (km/h)	V.ETR 460 (km/h)	V. VIRGIN (km/h)	V. TALGO (km/h)
1	F.V.	2.388	161	236	213	225	246	205	113
2	F.V.	3.180	214	314	284	300	327	274	150
3	F.T.	3.297	222	325	294	311	339	284	156
4	F.V.	3.789	255	374	338	357	390	326	179
5	F.T.	3.910	263	386	349	369	403	336	185
6	F.V.	4.019	271	396	359	379	414	346	190
7	F.T.	4.906	330	484	438	463	505	422	232
8	F.V.+T.	6.599	444	651	589	622	679	568	312
9	F.V.+T.	7.191	484	709	642	678	740	619	340
10	F.V.+T.	7.847	528	774	701	740	808	675	371
11	F.V.+T.	8.360	563	825	746	789	861	719	395
12	F.T.	8.595	579	848	767	811	885	740	407
13	F.T.	9.796	659	966	875	924	1009	843	463
14	F.V.+T.	10.171	686	1003	908	959	1047	875	481
15	F.T.	10.371	698	1023	926	978	1068	892	491
16	F.V.+T.	11.021	742	1087	984	1040	1135	948	521
17	F.T.	11.075	746	1092	989	1045	1140	953	524
18	F.V.+T.	11.800	794	1164	1054	1113	1215	1015	558
19	F.V.+T.	13.048	878	1287	1165	1231	1343	1123	617
20	F.V.+T.	13.194	888	1301	1178	1244	1358	1135	624

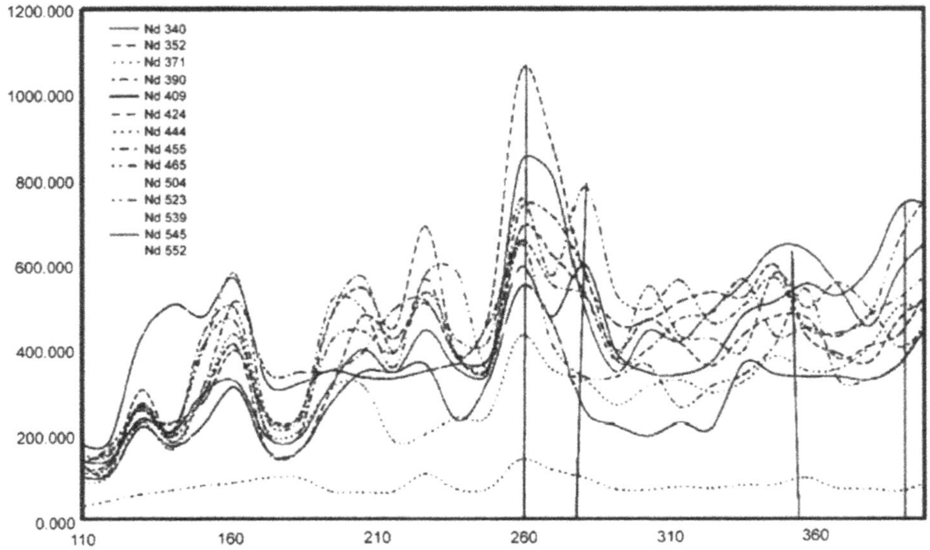

Figure 13. Dynamic response curves for ICE 2 from 110 to 400 km/h.

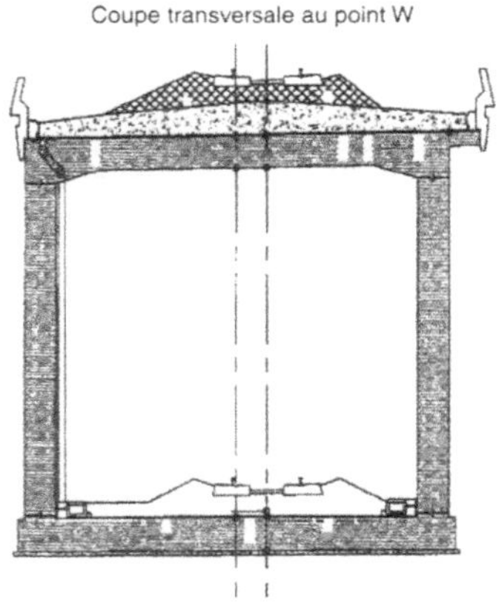

Figure 14a. Cross section of concrete portal.

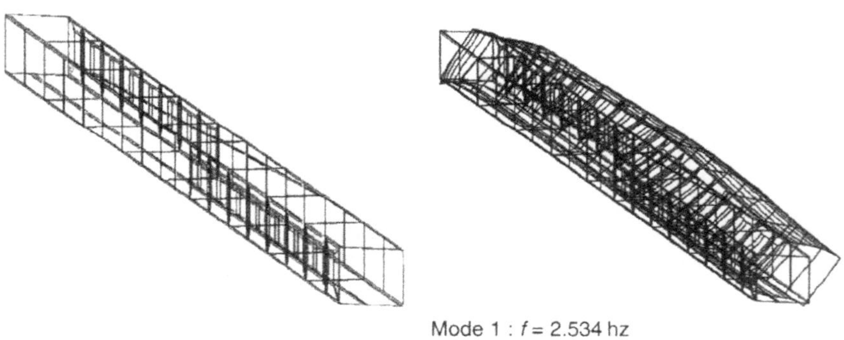

Figure 14b. FE Model and mode shapes of portal bridge.

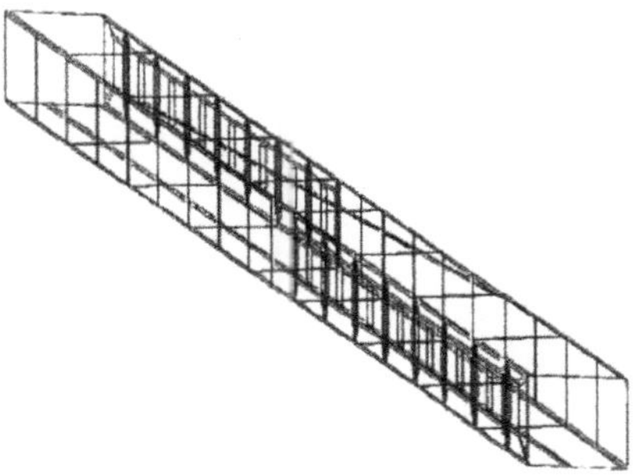

Mode 3: $f = 3.24$ hz

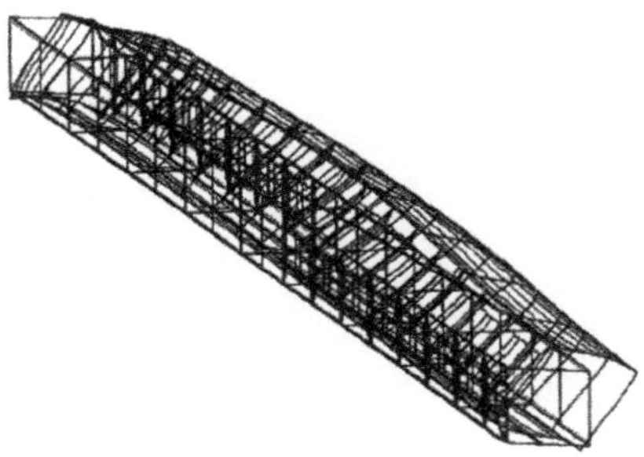

Mode 5: $f = 5.072$ hz

Figure 14b. (Continued)

Table 4. Critical speeds for the cracked concrete model.

MODE	Freq (Hz)	V.A1 (km/h)	V.A2 (km/h)	V.A3 (km/h)	V.A4 (km/h)	V.A5 (km/h)	V.A6 (km/h)	V.A7 (km/h)	V.A8 (km/h)	V.A9 (km/h)	V.A10 (km/h)
1	2.534	159	166	174	182	190	197	205	212	219	227
2	2.577	161	169	177	185	193	201	209	216	223	231
3	3.242	203	212	223	232	243	252	263	271	280	291
4	4.017	251	263	277	288	301	313	325	336	348	360
5	6.072	317	332	349	364	380	395	411	425	439	455
6	6.226	390	408	429	446	466	484	504	521	539	558
7	7.764	486	508	534	557	581	604	629	650	672	696
8	7.960	498	521	548	571	596	620	645	666	689	714
9	8.155	511	534	561	585	611	635	661	683	705	731
10	8.299	520	543	571	595	621	646	672	695	718	744
11	8.805	551	577	606	631	659	685	713	737	762	789
12	9.408	589	616	648	675	704	732	762	787	814	843
13	9.529	597	624	656	683	714	742	772	798	824	854
14	9.629	603	631	663	691	721	749	780	806	833	863
15	9.894	619	648	681	710	741	770	801	828	856	887
16	10.651	667	697	733	764	798	829	863	891	921	955
17	11.088	694	726	763	795	830	863	898	928	959	994
18	11.452	717	750	788	821	858	891	928	959	991	1027
19	11.545	723	756	795	826	864	899	935	966	999	1035
20	12.082	756	791	832	866	905	940	979	1011	1045	1083

7 CONCRETE PORTAL FRAME

This structure forms part of a railway interchange. The closed box portal has a straight alignment with a highly skew upper track (Fig. 14).

REFERENCES

OTUA: Bulletin Ponts Métalliques N° 19.

Ph. Ramondenc, L. Dielemann, W. Hoorpah. 2005. *Concevoir et réaliser les ponts métallliques autoancrés bow-string – Comportement sous charges ferroviaires et Comportement des véhicules.* Ponts Formation Edition Oct.

Le Bailly G. 2004. *Les ouvrages métalliques et mixtes de la LGV Est.* JT CETE-OTUA-RFF, Oct.

Friot D. 2004. *Les viaducs de Billy-le Grand et de Bussy-le-Chateau sur la LGV Est Européenne.* Steelbridge, OTUA.

Abi Nader I. 2005. *SDM 12 92 Inversion de V1 et V2 – Calcul dynamique*, IOA.

Hoorpah, W. 1997. *Contribution to the numerical analysis of dynamic behaviour of railway bridges.* Doctoral thesis UTC.

CHAPTER 10

Recent advances in the understanding of bridge dynamic behaviour on the West Coast Main Line Route Modernisation Project

P. Norris
Mott MacDonald, Croydon, UK

ABSTRACT: The European operation of new generation trains with equally distributed powered axles throughout the train has identified the potential for excessive vibration/resonance within bridge structures, leading to the potential for serious effects on structures and/or trackform, (i.e.) overstressing of structural elements and/or ballast instability. With this knowledge in mind, Network Rail commissioned a series of studies to investigate the likely magnitude and consequences of such a phenomenon, as part of the West Coast Main Line Upgrade. The initial phase of the project confirmed there was a potential problem and more importantly that there was a lack of knowledge regarding the behaviour of the dynamic system (bridge/trackform/vehicle) and the consequences of increased dynamic responses for typical UK bridge stock. The project has included shake table testing of track, site monitoring of some 150 structures under current rail traffic conditions to fully capture the dynamic responses and bridge specific analyses for up to 350 structures, complemented by a risk based assessment of a further 2000 structures, all with a view to achieving route clearance for new higher speed Class 390 passenger trains.

1 THE ISSUE

1.1 *Background*

The response of a bridge under the passage of the Class 390 trains at various operating speeds is determined by the structural form and characteristics of the structure. In general terms to ensure satisfactory performance this requires the structures to have:

- Adequate strength to sustain resultant dynamic load effects.
- Adequate stiffness to ensure deformations comply with current track safety acceptance criteria.
- An appropriate dynamic response to ensure that resultant vibrations (acceleration and displacements) are not excessive in order to avoid the risk of ballast instability.

The three main parameters, which influence the above resultant effects generated by trains at high speeds, are:

- Vehicle operating speeds.
- Vehicle/train Characteristics.
- Characteristics of the bridge which influence the magnitude of the dynamic response.

Existing vehicles/trains have been operating, to the best of our knowledge, at speeds up to 200 kph on existing UK bridge stock without concerns regarding:

- The strength of the bridges (provided adequate assessed route availability (RA) strength [UK equivalent of UIC Leaflet 700 line classification system] ratings have been calculated using conventional British Rail and Railtrack assessments techniques).
- Their dynamic response with respect to ballast stability.

The current conventional assessment criteria are based upon studies carried out by the UIC in the late 1970's. The risk of greater dynamic stresses was recognised and British Rail introduced a nominal 1.3 enhancement to dynamic factors for speeds in excess of 160 kph. However, more recent ERRI work indicates that such an approach may not always be sufficient for predicting dynamic bending stresses and fails to address critical bridge deck acceleration criteria. The introduction of new generation 'high speed trains' with their novel axle configuration, including the Class 390, requires careful consideration of their resultant operational effects on the existing UK bridge stock (including the effects of dynamics/resonance). This is due to their new axle loads and associated patterns/spacing, i.e. regularly spaced heavier axles.

Whilst generally it is understood/recognised that operating speeds and axle loads/patterns influence resultant dynamic load effects, it is not widely understood nor straightforward to predict what the dynamic effects (including those of resonance) are. These dynamic effects are set up in individual structures as a result of the complex interaction within a total dynamic system comprising: speed, axle loads/spacings and the individual dynamic characteristics of a particular structure and the track it supports.

Work undertaken on the West Coast Route Modernisation Project and elsewhere in Europe has highlighted shortfalls in the scope of current assessment and design standards in particular with respect to deck acceleration considerations and resonance effects. Studies have shown that whilst dynamic effects generally increase with speed, significant peaks and troughs in resultant effects occur throughout the speed range and are more likely to be critical as speeds approach 200 kph and above.

The studies undertaken have shown that the behaviour of the total dynamic system comprising of a train, the track and an individual structure is highly dependent upon the characteristics of an individual train. Where resonance effects occur, the use of a simple route availability or line classification type label to describe the effective weight of a train becomes invalid and two trains with the same RA number/line category can have very different dynamic effects. As a result, clearing a route for one specific train within a specified speed range may not guarantee satisfactory operation at similar speeds for another train. For each individual vehicle/train peak (critical) dynamic effects occur at different speeds within any speed range being considered and the magnitude of these peak effects varies for each individual structure and individual element within the structure.

In particular, the move from conventional power cars with heavy axles and repetitive lightweight coaches to trains such as the Class 390 with distributed traction systems and associated increased axle loads throughout the train represents a potentially more aggressive dynamic loading for bridges. It is therefore not sufficient to introduce new vehicles/trains simply on the basis that we have been successfully operating different existing vehicles/trains at similar speeds.

In recognition of the above, it was necessary to develop a route clearance strategy which ensured that the structures carrying the Class 390 routes have adequate strength to sustain the resultant dynamic load effects. Furthermore it was necessary to ensure that the resultant vibrations (accelerations/displacements) of the structures were not excessive in order to avoid the risk of ballast instability.

1.2 *Identification of hazards/risks*

Risks fall into two main areas, those associated with the bridge structure and those concerning the track form on the bridge structure. These risks are briefly discussed below:

– *Risks associated with the bridge structures.*

The main risks are with respect to the bridge structures strength and deflection. The risks relate to possible increased load effects resulting from the dynamic response of the whole system i.e. the bridge /track/train interaction.

The bridges are cleared adopting the following procedure:

- Where the results of a conventional broad assessment are published/available, structures will be cleared on the basis that the modified RA number for Class 390 (which will incorporate a

conservative adjustment for novel dynamic effects) is below the bridge specific rating being considered.
- Where assessments are not available, comparisons between real vehicle load effects for current traffic (HST/Freight) will be compared with real load effects of the Class 390 (including novel dynamic effects).

Where structures have insufficient strength based upon the above the Class 390 trains are restricted to a speed at which the safety of the structure can be ensured until such time as physical strengthening measures can be designed and installed.

– *Risks associated with the trackform on the bridge structure.*

The two main risks identified are:

- The loss of vertical support to the track, potentially resulting in changes in alignment, track twist, changes in load distribution, loss of wheel contact, rail fracture and/or the inability to sustain vertical load arising from ballast instability.
- The loss of lateral support to the track potentially resulting in, changes in alignment, track buckling and the inability to sustain lateral loads arising from ballast instability.

To examine ballast instability, checks are carried out to identify structures potentially at risk on the basis of generic bridge type/families (structure category) which have been derived utilising bridge specific dynamic analysis and parametric studies. Where potentially 'at risk structures' are identified control measures are implemented prior to, high speed, operation of the Class 390 to ensure that safe operation is maintained.

2 PROJECT APPROACH

The undertaking of this project has enabled a more complete quantification of bridge dynamic behaviour pertinent to the UK rail network and specifically the West Coast Main Line to be made, enabling the route to be cleared for the Class 390 tilting train operation. A structured approach to the task was adopted including:

- Development of a risk based strategy taking into account generic structural forms to identify high and low risk structures.
- Site monitoring and interpretation of bridge dynamic response under existing passenger stock.
- Strength and acceleration checks for the Class 390 train.
- Fundamental research using large scale shake table testing and small scale ballast tests to develop acceleration acceptance criteria applicable to the UK bridge stock.
- Development of generic short and long term physical mitigation measures.
- Verification testing of cleared and strengthened or replacement bridge structures under Class 390 operation.

For the purposes of this paper the issue of strength is not considered further. The development of techniques and procedures to quantify the dynamic response in terms of acceleration and displacement are covered in greater detail in the following sections.

3 INITIAL QUANTIFICATION OF THE SIZE OF THE POTENTIAL RISK

The West Coast Main Line (WCML) links London with Scotland, via the Midlands and North West England, serving the main conurbations of Manchester, Birmingham, Liverpool and Glasgow. Built in the 1860's, the line is the UK's busiest mixed traffic rail route carrying 15 million passengers a year and 43% of Britain's rail freight. The total number of structures requiring clearance for the potential effect of bridge resonance was in excess of 2000. During the initial stages of the work it was necessary to substantiate the completeness of Network Rail's Bridge records and 'walk' the

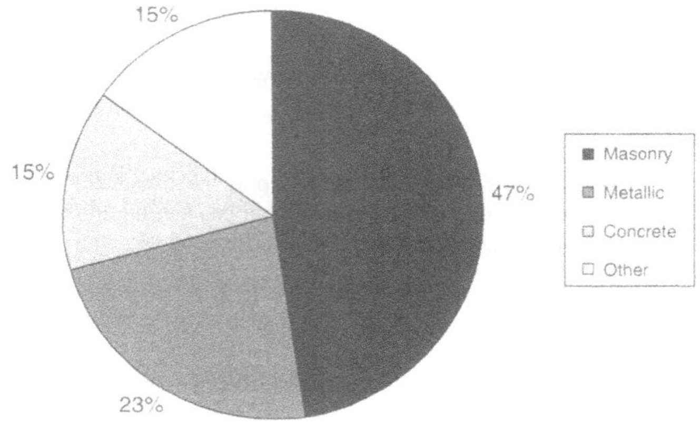

Figure 1. Composition of bridge population.

route to confirm each bridge, its general condition, structural form and its compliance with the information held.

Structures were then categorised into generic groups and their likely sensitivity to dynamic effects established, by taking into account the bridge type and the risks associated with the onset of ballast instability. Structures supporting lines with operating speeds below 144 kph, or perceived non-sensitive forms (masonry arches, culverts and backfilled structures) where not considered further, achieving the first stage of the route clearance.

The composition of the bridge population is illustrated in Figure 1. The metallic and concrete structure populations (perceived sensitive forms) comprised approximately 500 No. structures. These were further split into typical span lengths taking account of the different distinct forms. Representative structures within these groupings were then selected for structure specific monitoring and analysis on the basis that similar/twin structures could be cleared on a 'family' basis.

In this way the task was kept to manageable proportions; although by the end of the process up to 150 structures were monitored and 350 structures were analysed.

4 SITE MONITORING & DATA COLLECTION

Accurate prediction of future bridge deck behaviour under the operation of new generation trains requires bridge dynamic properties to be measured under existing traffic. The accuracy of any assessment based upon actual site data, will depend upon the accuracy of the data measured on site. Accordingly, careful consideration must be given to specifying and undertaking bridge monitoring.

Bridge monitoring encompasses the accurate recording of the existing bridge geometry, as well as monitoring deflection and acceleration of structural members under the passage of train loading.

4.1 *Key monitoring issues*

The following are considered to be the key monitoring issues:

(a) The mass of the structure has a significant effect upon its dynamic response. Accordingly, accurate and detailed measurements of ballast depths (across and along the structure) and structural elements are necessary.
(b) Span – bearing positions, type and expected bridge span should be carefully noted.
(c) The bridge should be monitored in such a way as to identify the main modes of vibration below the cut-off frequency of 50 Hz. (This frequency is set to capture the potentially safety-critical half-wavelengths of vibration over 1.5 m (i.e. approximately 2 sleeper spacings)). The main

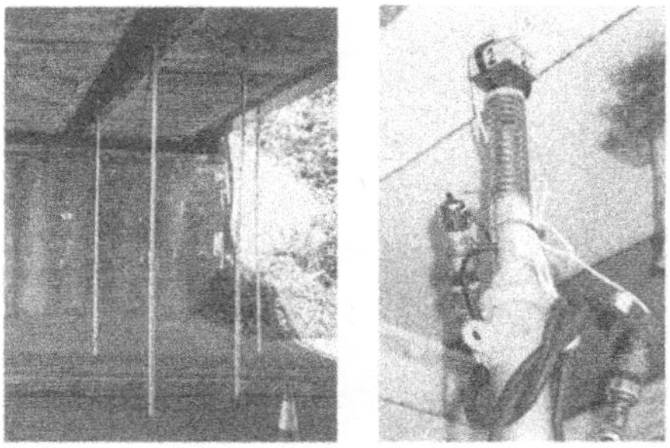

Figure 2. Accelerometers and deflection poles.

modes of vibration are typically: first longitudinal bending, first torsional and first transverse bending (deck). The presence of some of these main modes and other modes will obviously be dependent upon deck type.

(d) Sufficient data should be collected to provide the redundancy necessary to cross-check the data. At each point the displacement and acceleration should be measured to facilitate data validation. This will enable the displacement data to be double-differentiated to obtain the acceleration for comparison. (NB. It is difficult to double-integrate the acceleration to get the displacement.)

(e) The range of the load test instruments needs to be large enough to record the full behaviour, yet the resolution must also be sufficient. Although the displacements of short span bridges may be as low as 0.1 mm, the range of the instrument needs to be sufficiently large to facilitate site set-up. With regards to accelerometers, it should be noted that vibrations at frequencies above the cut-off threshold may cause over-ranging and clipping of the accelerometer traces. Subsequent filtering of the data may then mask the clipping leading to spurious results. The use of a large-range accelerometer will tend to reduce the accuracy of the measurements.

(f) Details of the train and its speed are necessary to carry out the validation of the analytical model. The speed of the train is readily obtained using a displacement time history graph to measure the time for a number of coaches to pass a fixed point. Site records must therefore record the type of coaches (or other vehicles) in each train. An indication of the relative loading of the carriages should be noted (i.e. presence of standing passengers, approximate percentage of occupied seats) where practical.

4.2 *Instrumentation*

For assessment work related to the WCML upgrade, 5 g or 10 g accelerometers, and 25 mm or 50 mm stroke linear potentiometric displacement transducers were utilised. Each potentiometer was mounted at the top of a pneumatically extending telescopic pole with a spring extending the shaft. The accelerometer was mounted at the top of the shaft so that it was held firmly against the bridge soffit. Figure 2 shows details of typical instrumentation used.

The instruments scanned at 3000 Hz with the digital trace then being passed through a 430 Hz anti-alias filter. (Aliasing is when higher frequency signals take on the identity of a lower frequency when converting analogue signals into digital data. It can cause large errors and reduce the validity of the data being collected. The anti-alias filter eliminates the higher frequency signals so as to avoid the aliasing phenomenon affecting the digital output).

The instrumentation positions were chosen to maximise the amount of useful data obtained from site. Generally, as a minimum, instruments where placed at mid and quarter spans. (The midspan readings give information on the main longitudinal mode and any torsional modes, whereas the quarterspan readings give information on higher longitudinal modes). Where transverse dynamic effects were anticipated (e.g. for half-through structures), the transverse members were also monitored (at midspan and their ends).

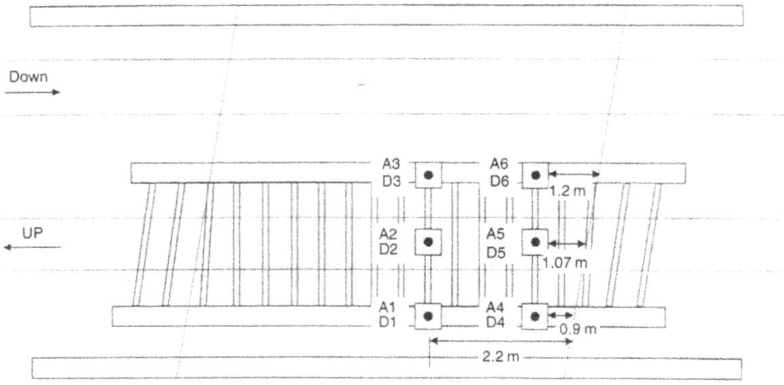

Figure 3. Typical instrumentation layout.

Figure 3 above shows a typical instrumentation layout. By not extending the monitoring past the centre of the span, it is usually possible to maintain traffic beneath the bridge. During the WCML work, the Up line was generally monitored as on this line the locomotive was usually at the rear of the train, which gives a larger excitation for the free vibration at the end of the measured trace.

4.3 *Initial data processing*

It is unlikely that all the traces from the various train events are suitable for the matching process. As such, primary data processing was undertaken to enable a smaller sample of traces to be chosen for further post-processing. This primary data processing consisted of filtering the data to the absolute frequency cut off (50 Hz), and reviewing the output on displacement and acceleration plots. The smaller sample was chosen by considering the train speed, type of train, and lack of irregularities on the trace, for each train event. It is prudent to include traces over a range of train speeds within the smaller sample. (Five train events were generally chosen for the smaller sample).

Digital output required from the initial data processing for subsequent detailed analysis included:

– Deflection vs time traces
– Acceleration vs time traces
} for each train event.

Some typical traces are included below in Section 5.

5 MODELLING, ANALYSIS AND MATCHING RESULTS

Accurate prediction of bridge dynamic behaviour is also highly dependent upon the modelling, analysis and matching processes adopted. Key considerations for each of these aspects are included in this section.

5.1 *Modelling*

Engineering judgment is required in deciding on the best type of finite element model. A simple model is often adequate and accurate enough to capture the bridge's main responses and can avoid unnecessarily complex analysis that is difficult to understand and prone to errors.

An initial line beam model of the bridge is an extremely useful tool to establish the basic parameters required in obtaining a good displacement match. Indeed, for 'simple' structures, e.g. deck type with low skew angles, line beams are often accurate enough to fully capture the bridge's responses in which case no further modelling is required. It is recommended that one should always start with a simple model and then move onto more complex models if required.

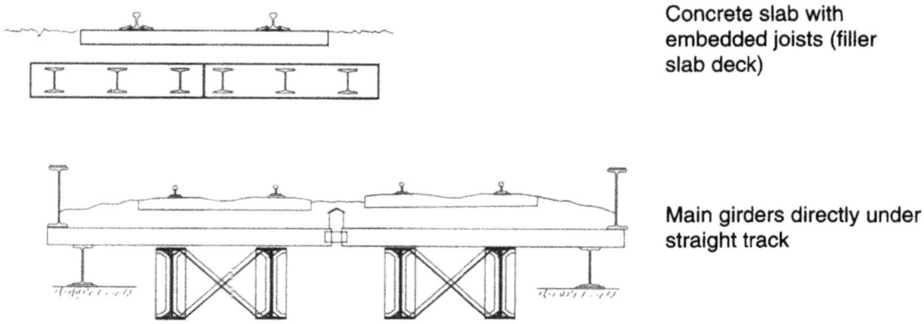

Figure 4. Structure types suitable for modelling using line beams.

5.2 *Types of models*

5.2.1 *Line beams*

Bridges comprised of longitudinal beams placed centrally under the track without skewed ends can be modelled with line beams. Line beams cannot be used when transverse modes are possible, i.e. structures where the track is supported on cross beams. Typical examples of bridge decks that can be modelled as line beams are shown below.

5.2.2 *Grillages*

Grillages are suitable for a wider range of structures as they include both longitudinal and transverse modes. Zero torsion grillages are acceptable in situations where the torsional stiffnesses of the beams are too low to attract any significant twisting moments. Where the deck is a continuum, a 3D FE model is preferable. Consideration must be given to the elevation of elements in the deck. Out of plane member effects must be considered in section stiffness calculations in order to achieve the correct global stiffness properties. Typical examples are half-through structures with floors that can act compositely with the main girders where the neutral axis of main girders and floor are at different levels. The individual element properties can be used if the appropriate out of plane offsets are included in the model.

5.2.3 *3D finite element models*

Adoption of 3d FE models was relevant for complex bridges with unusual geometry, vibration sensitive floors and structures with a large number of modes. Typical examples are bridges with transverse spanning members, light weight floor plates, thin slabs and skewed ends. Figure 5 shows a typical 3D finite element model of a half-through structure.

In developing the model it is essential to carefully consider connection details in terms of their stiffness and competency to sustain any applied load effects.

5.3 *Methods of analysis*

A time history analysis of the transient train event must be undertaken in order to capture the peak acceleration and displacement events occurring during the passage of the train. The most common methods of analysis for time histories are either by direct integration or by modal superposition. In both methods the solutions are obtained by numerical integration techniques. Direct integration techniques can be very costly both in time and storage requirements.

Modal superposition techniques are the most suitable type of analysis for the number of different speeds and options that need to be investigated. This method has the advantage that the modes being considered can be limited to the relevant frequency range and calculations can be limited to points of interest and the most responsive positions on the bridge.

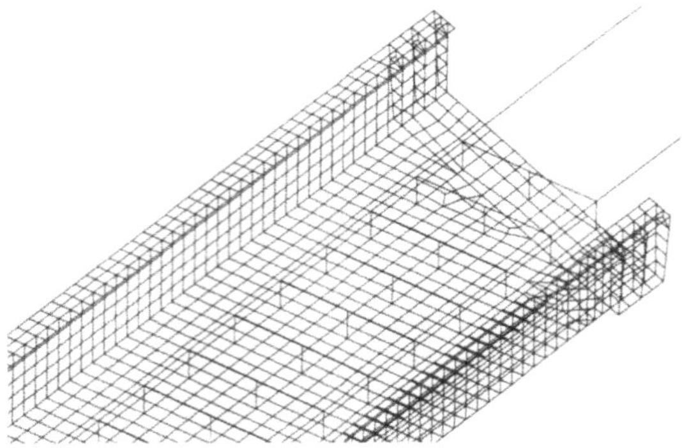

Figure 5. Example of typical 3D finite element model.

During this project, MABEL (Multi-modal Analysis of Bridges by Eigen Loading), a set of bespoke dynamic analysis programs utilizing a modal superposition technique, was developed, using Eigen values, which describe the vibrations of the bridge deck as a series of frequencies and mode shapes.

The use of the well-known Duhamel integral to solve a single degree of freedom transient response in the time domain at only a few critical positions on the bridge is the key to the highly efficient and computationally rapid solutions.

5.4 *Loading*

The distribution of load and the rate of loading are very important in obtaining accurate results. Point loads from the wheels are dispersed though the rail and ballast to the structure. The dispersion of the load is a function of the depth of the ballast and the relative stiffnesses of the rails, sleepers and ballast.

Both transverse and longitudinal distribution need to be considered. Load dispersion is achieved by adding a track model to the bridge. The track model needs to be wide enough and long enough to adequately distribute the loads. The rails should extend onto the embankment by at least five sleeper spacings. The rails should be longitudinally fixed at the extreme ends to simulate continuous welded rail and to avoid spurious sway modes. Rotational restraints at the start and end of the rail are required to ease the run on and simulate continuity of the rail.

The mass, size and stiffness of sleepers and rails must be included to derive full benefit from the track's load distribution effects. Pads must be modelled for directly fastened track.

Applying distributed loads can also simulate the dispersion effects. Depending on the application this can often be achieved by using multiple point loads. The distribution of load through sleepers and ballast can be assumed to be 50% on the directly loaded sleeper and 25% on the two adjacent sleepers.

5.5 *Time steps*

The length of the time steps used in analysis needs to be considered with regard to the speed of the train, the analysis method being used and the frequencies being considered.

The total number of time steps required for the passage of a train needs to be at least ten times greater than the highest frequency times the duration of event being considered. This is to smoothly describe the motion of the bridge at the highest frequency under consideration. Accelerations are calculated from the rates of change in velocity and where there are too few steps to describe the high frequency motions this will lead to erroneous acceleration values.

Train speed must also be considered when determining suitable time steps for the numerical integration. The time step chosen influences the position of wheel loads along the track. If the time steps are too large, wheel load positions will be too widely spaced.

Bridges with separate closely spaced cross girders are subject to impact loading induced by the rate of loading from the train approaching in a direction normal to the girder. Unlike a longitudinal girder, which has, half the span in which to reach its peak modal force a cross girder can be almost instantaneously loaded at its maximum modal force position. This is because the load is instantaneously applied at mid span of the cross girder where its mode has the largest magnitude. With poor longitudinal distribution between cross girders there will be little spread of load along the line of the track. (Floors in British Rail Western Region boxes are a good example of poor longitudinal distribution where the distribution between ribs is reliant on rail stiffness and a flexible floor plate). Accordingly the time-steps need to be chosen to ensure that these effects are taken into account.

It is suggested that ten load positions between each cross girder are employed to control the rate of loading. As an example, a train travelling at 50 m/s across cross girders at 0.5 m centres will require $50/0.05 = 1000$ time steps/second.

The above comments are also applicable to finite element models with discretised members where too large a time step will erratically load the element.

5.6 *Matching*

In order to obtain good correlation between site data and the model, the bridge frequencies, stiffness, span, mass and damping must be accurately matched in the analytical model.

The bridge's natural frequencies are the most critical parameters required for obtaining a good match with site data, and this is normally the first step in the matching process. It should be noted that the fundamental modes contribute to the majority of the structure's displacement, and as such this first step involves matching the displacement traces.

Higher order modes should then be matched with the site data. Higher order modes contribute the higher accelerations and as such cannot be ignored during matching.

However it must be noted that no adjustments to parameters should be made without justification.

For matching purposes damping is based on observed values. It should be noted that damping is an extremely complex parameter which required further fundamental research to fully define its contribution to bridge deck behaviour.

In practice, structures usually exhibit greater stiffnesses than those derived from the properties used for normal design purposes. For concrete sections it is normally acceptable to use the full concrete cross section. No allowance for cracking, or the use of cracked concrete section properties is normally necessary. Transformed section properties should be used where significant areas of reinforcement exist or the sections contain embedded steel sections or rails.

Pre-stressed concrete members must include the tendon stiffening effects. Site measurements have shown considerable increases in stiffness with the presence of pre-stressing. EI must be enhanced in order to match site measured displacements.

For riveted metallic girders the full sections including all angles and plates used for connecting webs and flanges should be used to determine dynamic effects (acceleration, displacements and forces).

Where the centre of the bearings are not well defined the span may need to be adjusted to improve the match. Large bearing plates are a typical example where the centre of pressure may not be at the assumed location. In practice the girders may be rotating about the front edge of the plate. A damaged cill may mean that the bearing centre has moved. Bearing support points can lie anywhere within the length of the bearing plate and as such, the actual span may differ from the theoretical span.

It is likely that the site data traces will include vibrations due to train loads, track roughness and track/vehicle defects, and it is important to differentiate between them. The average response derived from a set of repeating coaches is best for identifying responses without vehicle defects.

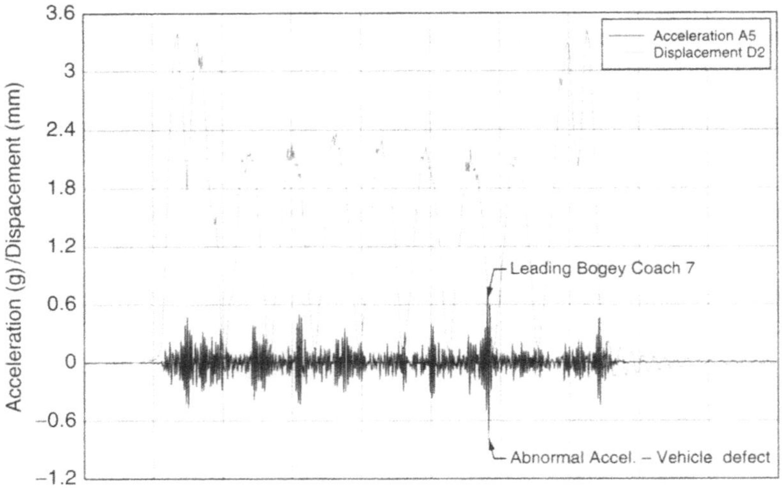

Figure 6. Plot of the variation in responses from coach loading.

All the coaches should induce similar acceleration responses, therefore an isolated abnormally high acceleration response will indicate a possible wheel defect for a particular coach. Variations in axle loading from differences in payload can be expected in the measured responses. Average loads from a typical coach with a normal passenger compliment should be used in the correlation, i.e. between tare and crush loading. Load data from rolling stock registers may be too high as it often allows for crush and abnormal loading scenarios.

Figure 6 is a plot of bridge displacement and acceleration responses to the passage of an Intercity train with a Class 43 diesel locomotive at each end and eight Mk III coaches. In the plot the response from coach 7 (higher than normal acceleration) can be attributed to a wheel defect. The acceleration responses from the coaches in this test are abnormally irregular indicating differences in loads and behaviour between the coaches. The differences in coach payloads are evident in the different levels of deflection between each coach.

Loads from electric locomotives are not subject to payload and fuel fluctuations, which make them more reliable for checking displacement responses. On the other hand locomotive induced accelerations are often quite noisy, as they tend to have stiffer suspensions and are more likely to suffer from wheel flats.

Obtaining a good overall displacement match will not automatically provide a good acceleration match. High order modes have small displacements, which introduce low amplitude high frequency detail into the signal. In Figure 7 calculated results for deflection and acceleration have been overlaid on test results. It can be seen that the overall displacements are closely matched, however in an enlarged more detailed examination the high frequency displacements are not apparent in the calculated displacements as shown in Figure 8.

This effect can be seen in the low amplitude waveform that is missing from the calculated results. The superposition of these high frequency displacements on the larger displacement has the effect of stopping and starting the motion. Bearing in mind that the gradient of the line represents velocity, a change in direction implies that the structure decelerates stops and then accelerates again to catch up to where it would have been. Examining the peaks between 0.45 and 0.75 seconds, eleven changes in direction and two shoulders can be seen in the test data. This implies that the structure slows down and increases speed twice as well as stopping and starting eleven times during this period. In contrast the calculated results only show the structure stopping five times. These abrupt changes in direction are the cause of high acceleration levels for short periods of time.

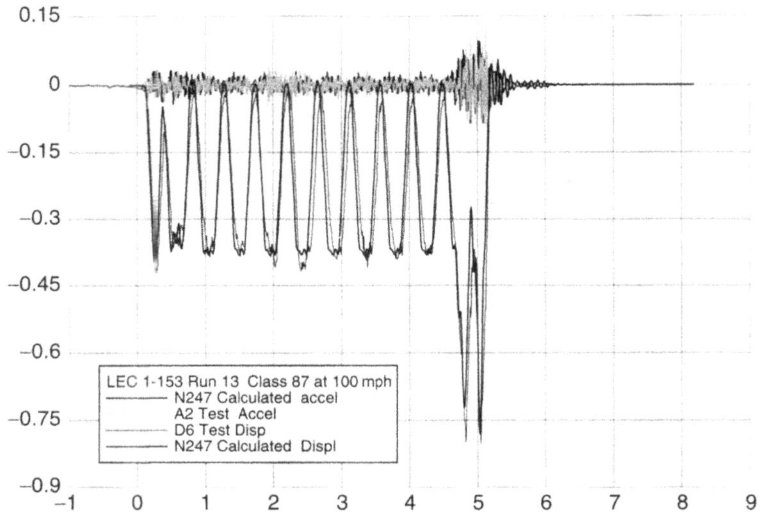

Figure 7. Typical acceleration and displacement match for a class 87 intercity train.

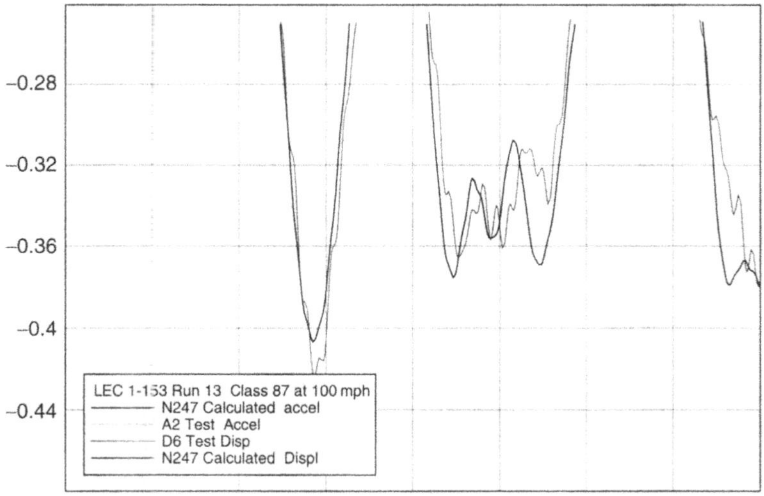

Figure 8. Enlarged detail of Figure 7 trace showing missing high frequency components in the calculated trace.

Examining the frequency content in the two signals will show that the calculations have only one frequency of 10 Hz, whereas the test data has many frequencies from 10 Hz up to 50 Hz.

Filtering the test data to 10 Hz will produce an almost perfect match which underlines the importance of selecting the correct cut off frequency and filter as well as identifying the source of the vibration. In many cases these small higher frequency vibrations are the result of train induced vibrations emanating from vertical vehicle motions due to track and/or wheel defects and are not from the bridge's natural vibrations.

Selecting two similar trains at different speeds provides the data to identify the vibrations from track and/or wheel defects. The vibrations from track corrugations are speed dependent, as opposed

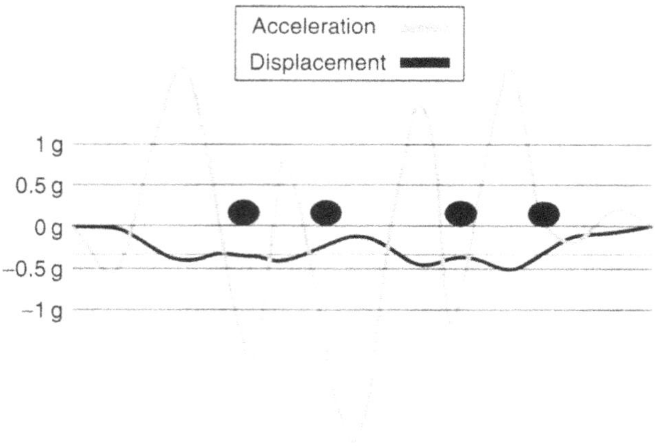

Figure 9. Section along the rails showing acceleration and displacement waveforms.

to the bridge's frequencies which remain constant. By scaling the time data for one of the trains the speeds of the two trains can be matched. The response spectra for the two trains can then be compared.

Matching the frequencies, mass, stiffnesses, damping etc will be an iterative process. An acceptable match is considered to have been reached when the differences between the site-measured peak accelerations and displacements and the analytical values are within the following limits:

Deflections ±5%
Acceleration ±20%

These limits assume that 'abnormal' effects due to track or vehicle defects, and/or excessive track roughness have either been filtered out of the site data, or included within the analytical model.

5.7 *Improved visualisation techniques*

In order to understand the dynamic bridge response, visualisation programs were written. One of these was used to show an animated graph of the displacement and acceleration along the deck. A second program showed an animated contour plot of the deck. From this, the extent of the regions of high acceleration could be seen. These visualisations were very useful in understanding and explaining the behaviour of the deck and consequences for the ballast.

The calculated accelerations and displacements of the floor directly under the rails at an instant in time for a short span high frequency bridge with a light weight floor are shown in Figure 9. The deflections directly under the wheels can clearly be seen. The short wave length deflections in the floor create the very high short wave length accelerations.

A snapshot of the acceleration contours for a similar short span floor with complex vibration patterns is shown in Figure 10. The bridge consists of two longitudinal main girders with cross girders supporting a flexible floor. The transverse cross girders, which are stiffening the floor and subjected to lower rates of acceleration, can clearly be seen in the figure.

6 ACCELERATION LIMITS FOR BALLASTED TRACK

Based on published best design practice in Europe a limit for bridge deck accelerations of 0.35 g was set for ballasted trackform (ENV 1991-3-2000, Annex H 3.1.2.1) and this was also initially

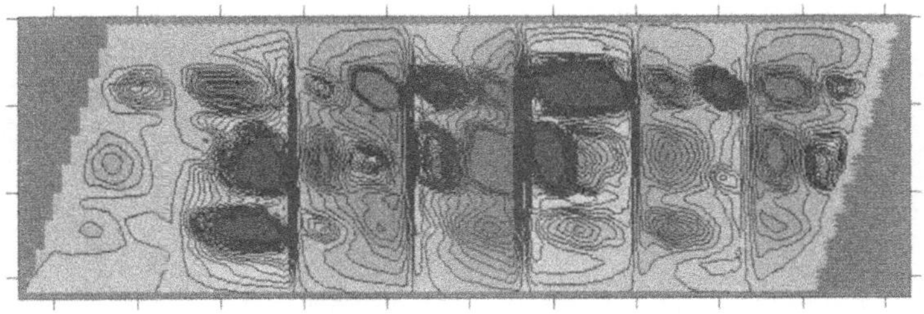

Figure 10. Single frame of animated acceleration wave plot showing localised high accelerations. The red areas indicate accelerations exceeding 1 g.

adopted for assessment purposes. However, these studies did not consider bridge deck vibration wavelengths/frequencies commonly found in typical UK rail bridges. This was considered potentially onerous as the amplitude/duration of vibration may provide alternative limits which safeguard against the onset of ballast instability and furthermore ballast containment (self or external) may limit the effects of ballast instability.

The objective of setting an appropriate bridge deck acceleration limit is to avoid safety critical changes in ballast behaviour. However, if the limit is unnecessarily onerous, this may result in large numbers of existing structures requiring precautionary mitigation measures that are unnecessary. It was considered that a greater knowledge of ballast behaviour under dynamic excitation was required.

6.1 *Quantification of the permissible acceleration threshold*

Fundamental support for the original 0.35 g new build acceptance criteria was obtained from shake table testing. A further shake table test programme was carried out at Bund esanstalt für Material forschung und prüfung (BAM) to provide additional data under controlled vibration and geometrical constraints. It involved sinusoidal testing of UK ballast within the frequency range 5 to 50 Hz and acceleration levels of 0.25 to 2 g. Due to operational constraints, the larger accelerations at higher frequencies could not be investigated.

General instrumentation arrangements included the measurements of acceleration at various positions in the ballast and on the rail grid and permanent settlement of the rail grid. Coloured ballast particles at various positions on the top surface and displacement gauges positioned in the ballast have also been introduced to provide quantification of ballast deformation within the ballast mass. Details of the test rig and the instrumentation layout are shown in Figure 11.

The objectives of the testing was to establish:

- the influence of the magnitude of acceleration on ballast instability with a view to deriving limits below which the ballast is not adversely affected (Loaded and unloaded sleepers)
- whether the duration of vibration is significant (short and long term cycling and impulse)
- the significance of ballast lateral confinement (ballast shoulders and varying edge distances to adjacent structural members).

6.2 *Key findings*

Shake table testing allowed trackform in a loaded and unloaded state to be subject to controlled frequency-acceleration conditions. Testing was carried out using ballast, rail and concrete sleepers to Network Rail specification, as incorporated into the WCML.

The tests enabled the potential ballast instability mechanism to be established. This involves dilation of the ballast leading to an increased rate of settlement and heave between the sleepers. This phenomenon was seen to manifest between 0.9 to 1.0 g under the wide range of shake table

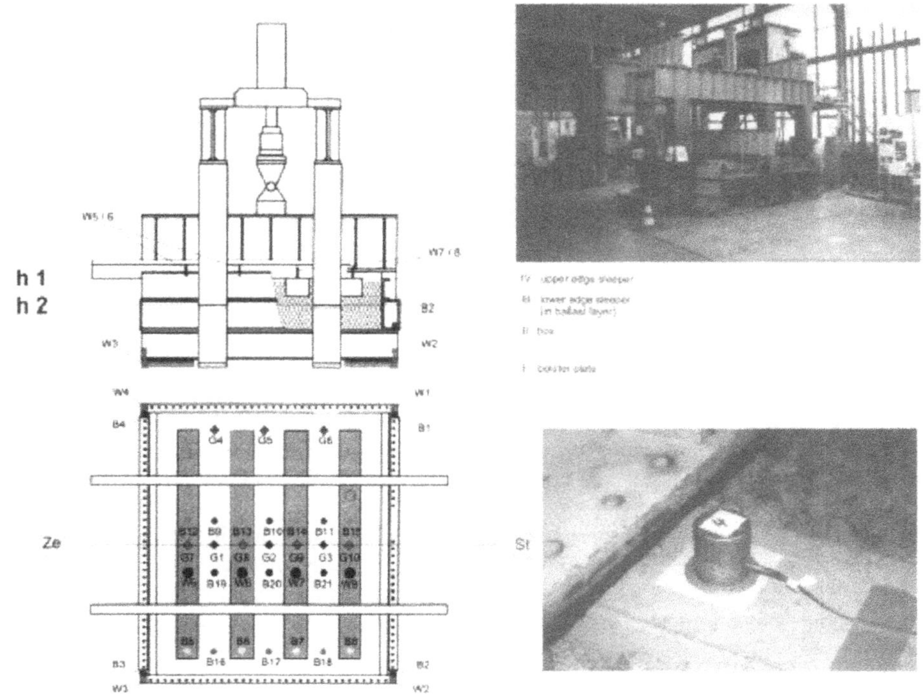

Figure 11. Details of test rig (left), shake table set-up (top-right), sleeper accelerometer (bottom-right).

tests, which included different geometrical, loading and duration variations. 0.9 g has been taken as the threshold acceleration at which ballast instability is assumed to occur as in Figure 12. This is consistent with the observations in the ERRI D214 work. None of the shake table experiments showed what could be termed safety critical behaviour within the limits of the testing.

The shake table test programme was used to evaluate amplification or transfer functions, i.e. increase in acceleration between base and top surface of the ballast as a function of frequency. These enabled the resonant frequency of the unloaded and loaded trackform to be established at 35 Hz and 50–60 Hz respectively. The amplification at resonant frequency in the tests was shown to be in the order of three, (i.e. an input base acceleration of say 0.5 g would produce 1.5 g on the top surface). This amplification of acceleration is a maximum at the resonant frequency of the trackform and reduces rapidly at frequencies lower or higher than the resonant frequency.

These values can be compared with the measurements of amplification and resonant frequency obtained from actual bridge sites. Unfiltered site data confirmed that the resonant frequency of trackform was approximately 100 Hz during the train passage with a similar peak amplification of three. For typical structures on WCML, the 0–50 Hz frequency range captures the dominant modes of vibration. Power Spectral Density (PSD) data obtained from site monitoring indicates that the energy within vibration at frequencies higher than 50 Hz is negligible. Because resonant frequency for the actual trackform is relatively high compared with the dominant bridge frequencies (i.e. less than 50 Hz) the amplification which occurs is negligible. This is important since trackform instability is likely to occur when the ballast below the track can dilate within a low stress environment. As amplification reduces the base acceleration has to increase for near zero stress to develop below the track. Hence, because the shake table tests induce higher amplification through the ballast than those that develop at actual bridge sites, the shake table tests are likely to give a conservative indication of ballast and track instability for typical UK bridge stock. The behaviour

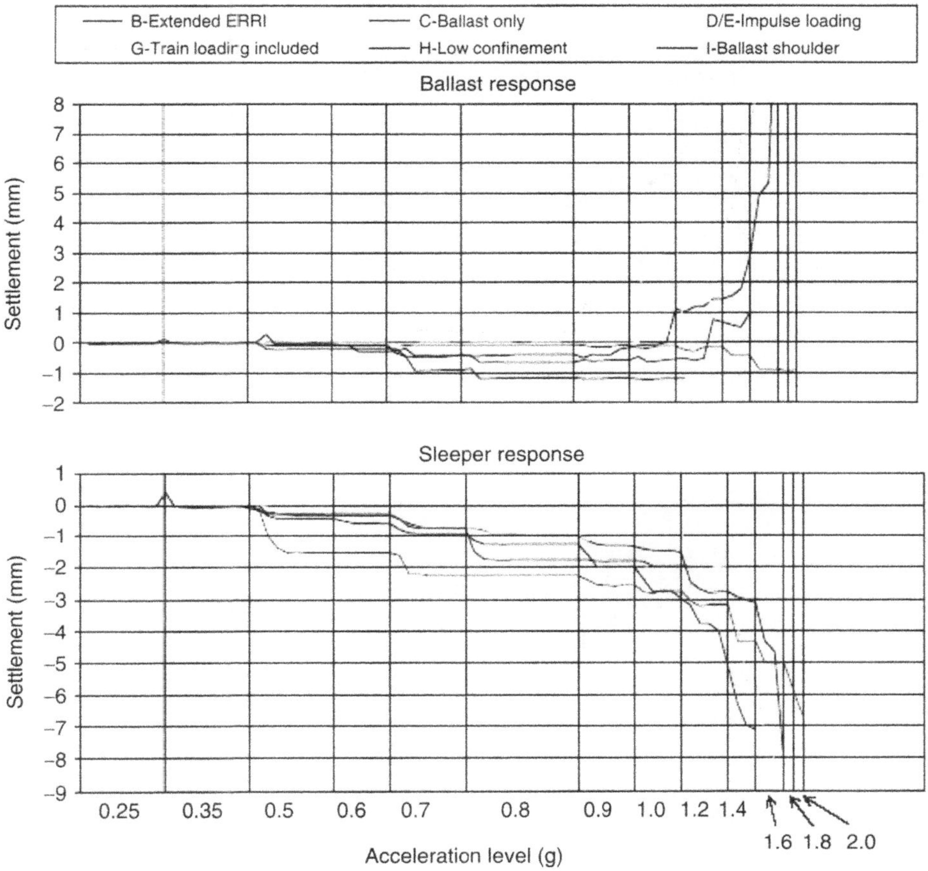

Figure 12. Summary of sleeper and ballast settlements during shake table test series.

for structures with low frequency – high amplitude responses and subject to resonance effects is considered to be more representative of the conditions in the shake table. The shake table testing did not facilitate the incorporation of a number of potentially important factors such as combined vertical and lateral loading, future changes to ballast depth and the effects of ballast degradation with time and changes to material type. These have by necessity, had to be estimated on the basis of existing knowledge.

6.3 *Quantification of the permissible extent of unacceptable acceleration*

A formal risk assessment was jointly undertaken with key Mott MacDonald and Network Rail, track, geotechnical and bridge dynamic specialists to establish the likelihood, cause, effects and consequences of the onset of ballast instability. The output of the workshop was that track was considered capable of accommodating transient contained localised instability, provided such zones were less than 2 sleeper spacing for vertical and lateral load effects. The contour plotting program used to establish the extent of peak accelerations arising during the passage of a train (section 5.7) confirmed that many short wavelength (high frequency) bridge elements illustrated localised peaks, which did not extend more than two sleepers in extent.

For illustrative purposes the plots of a low frequency concrete structure and a half through metallic box structure with a shallow orthotropic floor are presented in Figure 13 and Figure 14.

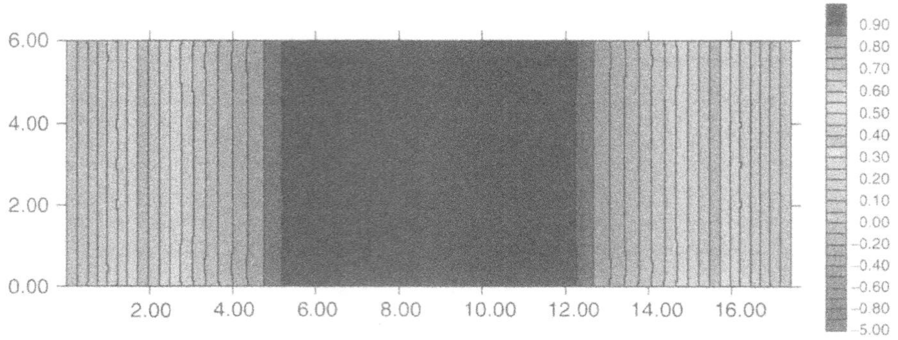

Figure 13. Low frequency concrete structure response to articulated train set.

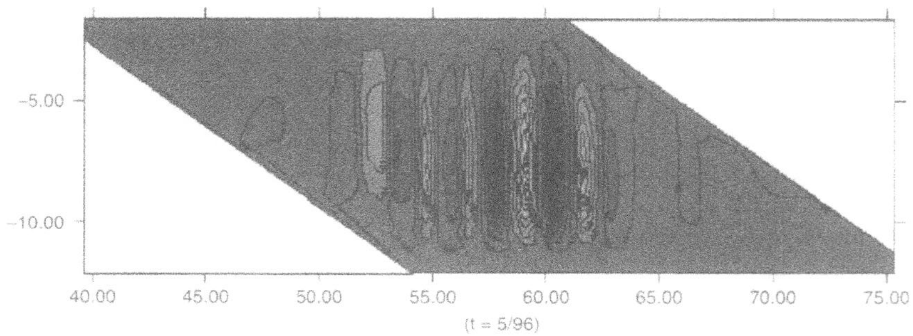

Figure 14. Orthotropic floor acceleration response to Class 390 tilting train with a conventional configuration and distributed traction.

The plot of the long span structure shows the effects of global deck action, i.e. single wave profile. It vibrates four times a second and subjects ballast, over a 5 m length, to zero gravity effects for approximately one tenth of a second during each cycle. By comparison localisation of high acceleration effects is clearly demonstrated in the high frequency orthotropic floor which has poor longitudinal load distribution which promotes moving wave type behaviour with the ripple effect subjecting the ballast to both positive and negative acceleration of similar magnitudes. For half the cycle, with positive accelerations, the ballast will be subjected to higher stresses making the material stronger. For the other half of the cycle, the ballast will have reduced vertical stresses giving rise to concern with regard to ballast stability. The stress cycles may also lead to longer-term maintenance issues. The loosening ballast zones are typically spaced at 2.5 m centres and extend over a width of 1.0 to 1.2 m. A second plot for this structure taken 20 milliseconds later shows that the areas previously subject to downward acceleration are now accelerating upwards. No growth in the zonal extent of the larger accelerations is evident from this work. The resultant safety implications between this structure and the low frequency structure are markedly different.

It is important to note that acceleration limits based on shake table testing, have been derived from ballast subject to equal acceleration at any instant in time over the full extent of the shake table. Generic structure types likely to be producing high accelerations on WCML (floor elements of halfthrough structures and short span bridges <5 m) exhibit localised potentially damaging acceleration events. The potential to increase the allowable acceleration above that used for global deck type behaviour exists.

6.4 *Acceptance criteria for dynamic assessment on WCML*

A deck acceleration limit of 0.5 g for the assessment of existing bridges was applied to the most susceptible bridge decks with ballasted track. These decks have:

- Long acceleration waves at a fundamental frequency such that the train forcing frequency (or multiple of it) may match it within the possible speed range. Examples of long acceleration waves are the first and second longitudinal modes with wavelengths of 2 × span and 1 × span respectively.
- Sufficiently low damping that resonance amplification may occur.

For less susceptible decks, an acceleration limit of 0.6 g was applied. Although these decks also have long acceleration waves and ballasted track, resonance will not occur because either:

- The forcing frequency cannot match the bridge fundamental frequencies, or
- Damping is sufficiently high that little resonance amplification occurs.

Higher acceleration limits up to 1.0 g can be accepted for the assessment of existing bridges for fully confined ballast with high frequency short acceleration wave lengths provided the following criteria can be satisfied:

- Half-wavelengths are short enough to not disturb more than two sleepers, i.e. half wavelengths must generally be less than 1.5 m.
- Vibrations are stochastic, i.e. isolated acceleration peaks occur under the passage of each vehicle.
- Vibrations are not cyclic, i.e. accelerations of similar magnitude are not repeated at equally spaced time intervals causing resonance in the ballast.

The acceleration wavelengths described above are calculated beneath the track where the train is most vulnerable to ballast destabilisation and loss of support.

7 MITIGATION SOLUTIONS

For the WCML project, programme considerations were of paramount importance and it was therefore necessary to develop both short term and long term mitigation solutions in parallel with the route clearance process, to provide flexibility to the planning process.

Key factors taken into account when formulating options include:

- Site survey, geotechnical surveys, material testing, infrastructure survey.
- Conventional structural assessment results.
- Primary constraints (i.e.) possessions kept to minimum, structural gauge, kinematic envelope and headroom requirements, financial considerations.
- Removal and reinstatement of trackwork, or track realignment.
- Statutory Undertakers plant.
- Environmental aspects.
- Health and Safety considerations.
- Rights of access (Land Ownership).
- Installation/erection.
- Planning consideration.
- Railway Inspectorate Approval.
- Requirements for monitoring.
- Review of existing records.
- Long term safety of the Operational Railway.
- Financial considerations.
- Possession/Temporary Speed Restrictions.
- Temporary works.

- SAT/Plant diversions.
- Disturbance to the existing structure.
- 'Traffic' management.
- Installation methods.
- Durability, Serviceability, Maintenance.

7.1 *Short term solutions*

7.1.1 *Ballast bonding*

Ballast bonding/gluing was used to stabilise ballast in and around sleepers to ensure that sleeper bearing was not compromised by movement of the stone due to dynamic excitation of the bridge deck. The intent was to create a blanket of bonded ballast to contain/restrain the stone below sleeper level.

The bonding agent was gravity fed to the top of the ballast. The rheology of the material was formulated for each specific use to allow appropriate penetration into the ballast and to give the required properties to the bonded stone. Generally, it bonds only the contact points of the aggregate.

Installation on site was relatively easy with the required treatment being capable of being carried out whilst trains were running. The main equipment consists of a compressor, pump and applicator lance, held by an operator. A realistic treatment rate of between 1 and 2 minutes per sleeper was achievable. The process is unaffected by damp weather conditions and is reasonably tolerant of dirty stone. It is not possible to tamp the track at glued locations and the track alignment cannot be altered without removing the glued stone. The bonded ballast can be broken out but requires pneumatic/electrical hand tools. Removal will break the stone rendering it unsuitable for replacement unless it is re-bonded.

The technique, if required, can be used as a stand-by mitigation process, with enhanced confidence that it can be effectively implemented within tight construction timescales.

7.1.2 *Additional rails*

Short span bridges, in the order of 5 metres, can be dynamically improved in the short term by the addition of track or steel sections fixed to the sleepers to improve longitudinal distribution of load. For example, an additional six rails attached to the sleepers could possibly give a 15–20% reduction in predicted accelerations. The rails would need to extend on to the approaches, requiring lengths of approximately 10 metres to be installed at each structure.

It was noted that many sleepers on the WCML bridges were likely to be up to 30 years old, and their condition variable. Generally timber sleepers have been used across the bridges, with concrete sleepers on approaches. Methods of fixing the additional stiffening rails to the different sleeper types required investigation, evaluation and testing. As a result of these factors, this option for short-term mitigation was never adopted in practice.

7.2 *Long term solutions*

7.2.1 *Stiffening solution*

Bridge CMP1/92 is a single span structure carrying two tracks. It consists of two main longitudinal box girders carrying a station platform, with cross girders, bulb T rail bearers and steel deck. The bridge spans 18.5 m and has a skew angle of 16°. The required line speed and design speed where 200 kph and 240 kph respectively. Predicted acceleration in the floor and cross girders were required to be reduced. Predicted accelerations in the main girders were found to be acceptable.

Stiffening the floor transversely by the introduction of additional members between existing cross girders was found to be ineffective. The adopted solution involved the introduction of a longitudinal stringer along the centre of the deck and stiffening the floor of each deck unit.

Details of the mitigation solution are presented in Figure 16 below.

All work was carried out below the bridge with no requirements for a temporary speed restriction.

Figure 15. Bridge CMP1/92.

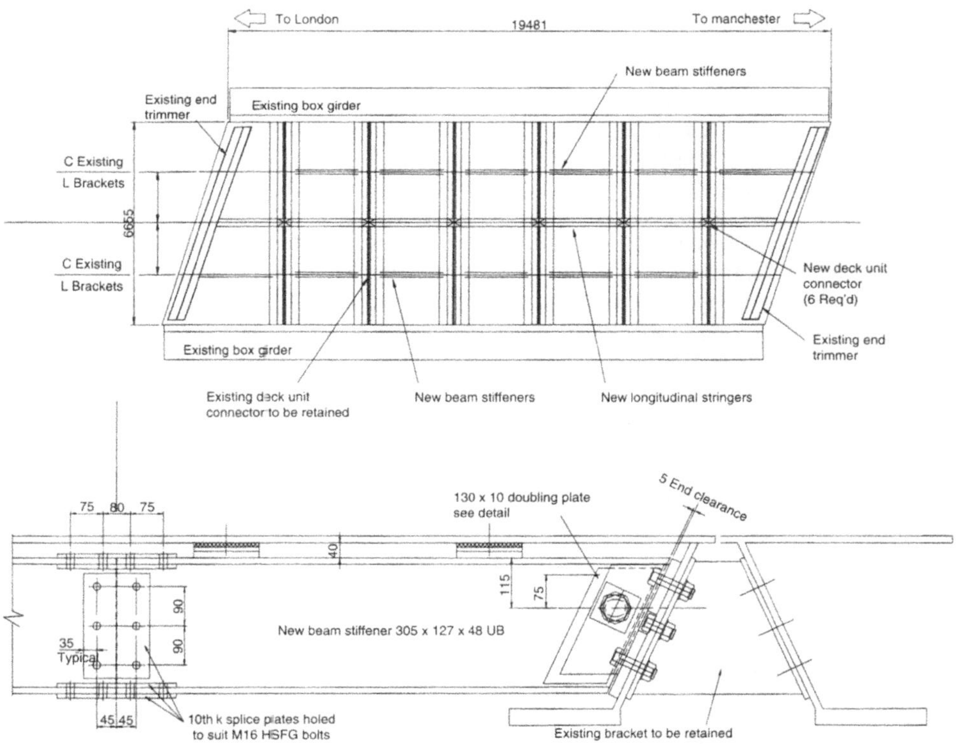

Figure 16. Bridge CMP1/92 mitigation solution.

7.2.2 *Low vibration bridge replacement design*

The route clearance process identified a number of superstructures where high-speed traffic may lead to excessive deck excitation with associated risk of ballast instability. Intricate and costly stiffening work was deemed not feasible for five such superstructures and replacement with a novel U-deck developed by Mott MacDonald was the client's preferred option.

Figure 17. U-deck replacement.

The replacement structures, have minimal construction depths and to comply with the stringent dynamic performance criteria imposed for running speeds above 200 kph, relatively high mass and stiffness is required. For these reasons the superstructure is formed of thick plates comprising a 150 mm thick steel deck slab, top flanges up to 200 mm thick and thick web plates that require minimal stiffening. A typical cross section is shown in Figure 17.

This configuration provides the overall stiffness required to avoid excessive dynamic effects, optimises construction depths and does not require a reduction in ballast depth or headroom. The plate thickness exceed those commonly used in short to medium span UK bridge construction so material specifications and welding procedures and details were developed alongside the detailed design. Particular attention was paid to the provision of main and duplicate load paths through connections with a view to achieving the security offered by the original U-Deck and traditional designs.

To date, spans from 7 m to 12 m and angles of skew up to 45° have been detailed and it has been possible to provide the required minimum 300 mm ballast with construction depths as little as 850 mm. The simplicity of the design has minimised fabrication time and the philosophy behind the U-type structures has been upheld by providing a low construction depth, robust structure that takes full advantage of maximising prefabrication.

8 VERIFICATION TESTING

The fundamental purpose of verification testing on the WCML was to:

i) Verify the original predictions used to confirm the acceptability of families of structures.
ii) Confirm that design intent had been achieved for the replacement and stiffened/strengthened structures (which were implemented as a consequence of the bridge resonance studies).
iii) To investigate the effect of track maintenance on bridge dynamic response.

A total of eight bridges were subject to verification testing during October 2004. Each bridge was monitored under existing traffic consisting of Intercity and Class 390 trains up to speeds of 200 kph.

Tests to verify the applicability of the modelling and prediction process used to clear the route were performed on Bridges LEC2/91, CMD1/50, and WCM1/181 which were chosen to represent

Figure 18. Bridge LEC2/91.

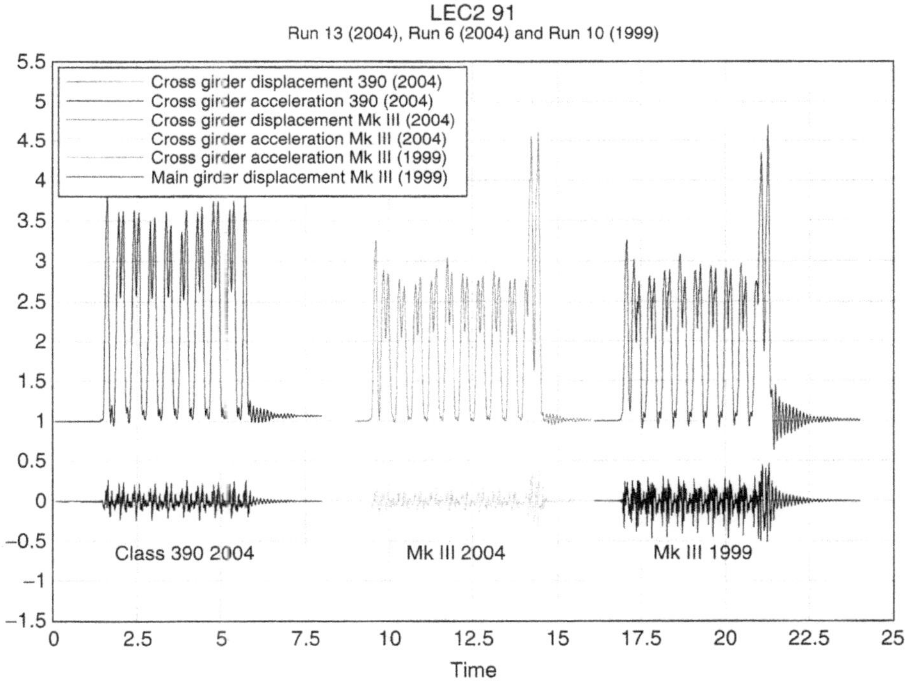

Figure 19. LEC2/91 Comparison of Mk III coaches in 1999 and 2004 and class 390 in 2004.

structure types that had a large population of similar structures. A selection of results for Bridge LEC 2/91 are included in Figure 18 to Figure 21. The actual dynamic behaviour was found to be less onerous than the prediction on which it was cleared. Comparison of the behaviour under MkIII coaches captured during the original and most recent monitoring exercises showed that the level of

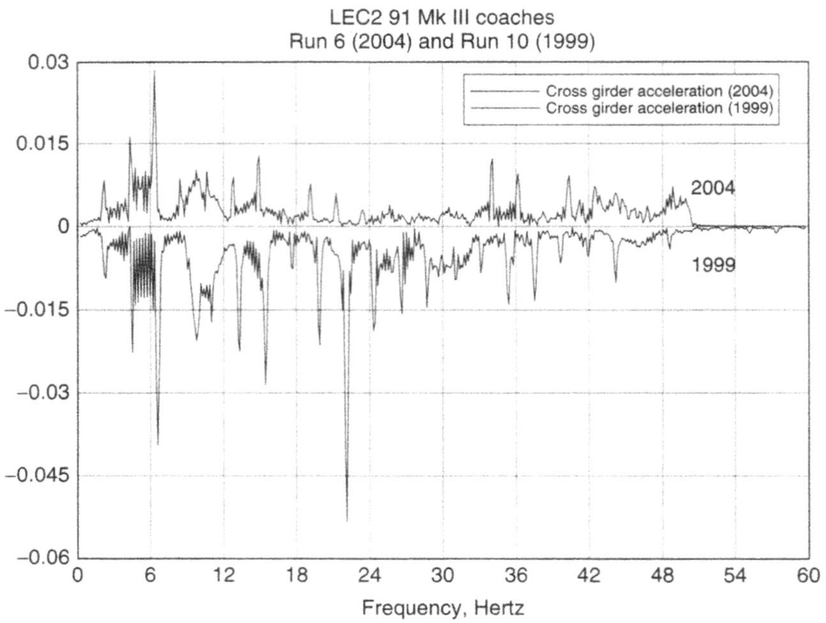

Figure 20. LEC2/91 comparison of frequency response 1999 and 2004.

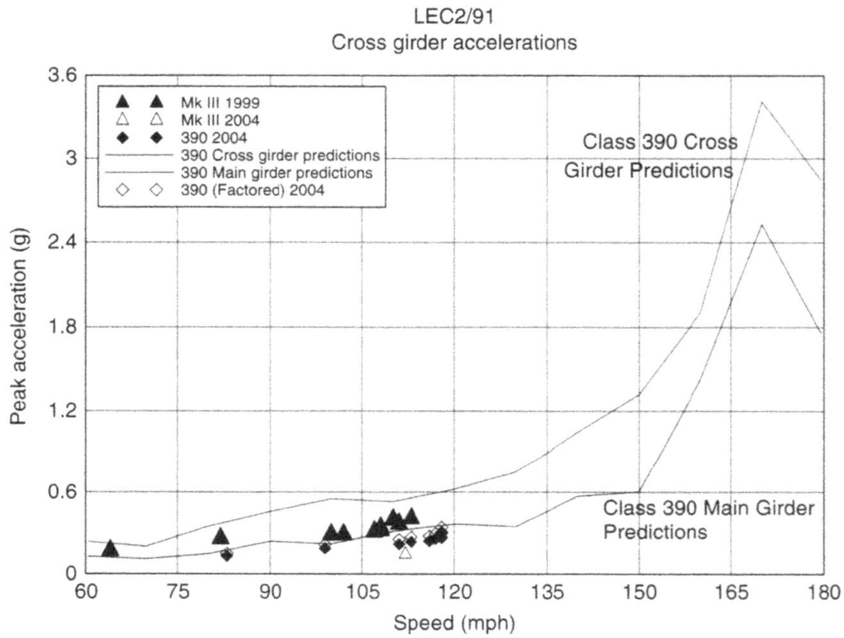

Figure 21. LEC2/91 predictions and measured results.

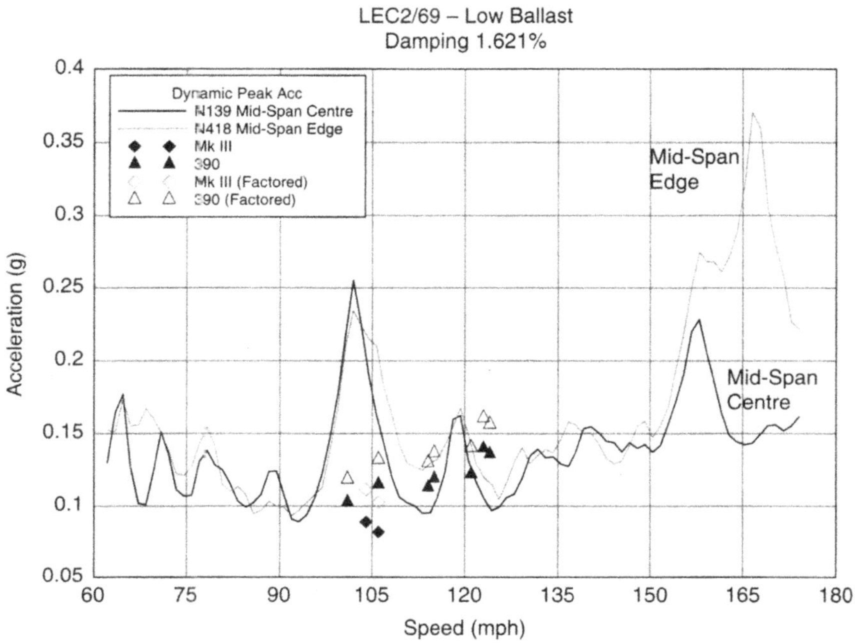

Figure 22. LEC2/69 predictions and measured results.

vibration had actually reduced. The most likely reason was considered to be improvements in the condition of the track which was part of the upgrade as suggested by the comparison of the response spectra which show a significant reduction in response for frequencies above 12 Hz. The original predictions are considered appropriate as future determinations in track condition will cause the dynamic effects to increase tending towards the predictions.

Tests to confirm the design intent for stiffened and replacement structures was achieved were carried out on LEC2/53 (Stiffened) and LEC1/100 and LEC2/69 Up ('A' Type replacement).

Testing of the replacement A type structures confirmed that design intent was achieved. Figure 22 shows the prediction and monitored results for Bridge LEC2/69. The peaky values in the prediction are due to the low damping (1.6%) used in accordance with the design code rather than the measured value of 5%.

Investigation of the effects of track condition on bridge dynamic behaviour and implications for maintenance on the WCML have confirmed significant improvement in the dynamic performance of short span structures when track condition is improved.

9 CONCLUSIONS

- A large data base of dynamic response data has been captured for UK bridge types which has subsequently been used to develop rapid route clearance tools for the introduction of other new generation trains.
- Shake Table testing has allowed trackform in a loaded and unloaded state to be subject to controlled frequency – acceleration conditions, leading to enhanced ballast acceleration acceptance criteria for the West Coast Main Line project.
- The development of bespoke software using modal superposition techniques that could be benchmarked against test data whilst providing rapid analysis to a very complex problem was

essential in gaining confidence in the accuracy of the solutions and in the understanding of the dynamic behaviour.

- It was adequately demonstrated, through close correlation with test data from thousands of train runs, that train/structure interaction effects and abnormal train and track defects could be accurately simulated without resorting to a fully interactive train/structure analysis using direct integration. Varying the transient modal forces as a function of the instantaneous vertical acceleration and unsprung mass together with modal superposition provided an accurate and very rapid solution to forced and natural vibrations for short span bridges with high frequencies subjected to both train and track defects.
- The development of graphics tools to visualise the results has not only improved our understanding of dynamic wave propagation but has allowed us to define the spatial extents of ballast instability and the zones of ballast constraint at any instant in time.
- Effective short and long term mitigation solutions have been developed for dynamically sensitive structures. For short term solutions, a form of ballast bonding has been shown to be cost effective and rapid to implement. For longer term mitigation, the option of retrospectively stiffening was shown to be difficult to achieve during rail possessions, for the majority of bridge types under consideration. The exception was light weight metal floors consisting of a grillage of beams and floor plates which can be stiffened from beneath. Structures requiring replacement were largely half-through bridges, the structural form being dictated by clearance and headroom requirements. A low vibration alternative has been developed capable of supporting 225 kph operation.
- Verification testing was carried out once Class 390 trains were operating at 200 kph to ensure that predictions were sound and no unforeseen behaviour was apparent. The results of this testing have confirmed that predictions were accurate and that the enhanced acceleration acceptance criteria is safe.
- The development of the techniques covered by this paper has enabled a large population of structures along the West Coast Main Line to be safety assured (for higher speeds of operation of the new Class 390 passenger trains) within a very tight project programme, as well as enhancing fundamental understanding of high frequency vibration in short span bridges which can be readily transferred to other routes.

ACKNOWLEDGEMENTS

The author would like to acknowledge the contributions of Network Rail HQ and The West Coast Route Modernisation Team and those organisations and individuals who have contributed to the knowledge presented in this paper. The views expressed are those of the author and not necessarily those of Network Rail, Mott MacDonald or any other organisation referred to.

REFERENCES

ERRI Committee. 1999. D214/RP9 Rail Bridges for Speeds >200 km/h. Final Report, Part A: Synthesis of the results of D214 Research, ERRI, Utrecht, 362 pp.

DD ENV 1991-3-2000 Annex H 2.1.2.1 – *Traffic Loads on Bridges*.

Author index

Subject index

For Product Safety Concerns and Information please contact our EU
representative GPSR@taylorandfrancis.com
Taylor & Francis Verlag GmbH, Kaufingerstraße 24, 80331 München, Germany

www.ingramcontent.com/pod-product-compliance
Ingram Content Group UK Ltd.
Pitfield, Milton Keynes, MK11 3LW, UK
UKHW052211190926
475702UK00006B/973

* 9 7 8 0 4 1 5 4 6 7 6 7 4 *